Helga Nowotny

In Search of
Usable Knowledge

Utilization Contexts and the Application of Knowledge

Campus Verlag · Frankfurt am Main
Westview Press · Boulder, Colorado

Printed with the financial support of the Austrian Federal Ministry for Science and Research

Library of Congress Cataloging-in-Publication Data

Nowotny, Helga.
In search of usable knowledge / by Helga Nowotny.
p. cm.
  ISBN 0-8133-7958-X
  1. Social sciences-Methodology. 2. Time-Social aspects.
3. Information technology–Social aspects. 4. Science-Social
aspects. I. Title.
H61.N68 1989   89-48579
300.72-dc20   CIP

CIP-Titelaufnahme der Deutschen Bibliothek

Nowotny, Helga:
In search of usable knowledge : utilization contexts and the
application of knowledge / Helga Nowotny. – Frankfurt am
Main : Campus Verlag ; Boulder, Colorado : Westview Press,
1990
  (Public policy and social welfare ; Vol. 3)
  ISBN 3-593-34245-6 (Campus Verlag)
  ISBN 0-8133-7958-X (Westview Press)
NE: GT

© 1990 by European Centre for Social Welfare Policy and Research, 1090 Vienna, Berggasse 17.
Composition: Werner L. Laudenbach. Printed by Interpress – Dabasi Nyomda, Hungary.

Dedicated to the European Centre,
my institutional home from 1974–1988,
and to my friends and colleagues.

Contents

Introduction

Having worked for almost fourteen years as Director of an international institution devoted to social policy and social welfare development in the European region of the United Nations, I have been extremely fortunate in drawing upon the ideas and efforts of others in almost everything that has been published by the European Centre during these years. Looking backwards in time, I feel that together with many friends and colleagues we have been able to contribute to the creation of a common discussion space in the field of social policy, crossing national borders, including those that run between countries situated in Eastern and Western Europe. All in all, it has been a very intense and fruitful period of my life – helping to found and build such a unique institution as the European Centre and leaving some marks, I hope, in terms of intellectual content and policy-relevance on different local as well as international utilization contexts. These are contexts that bind people together, their aspirations and ideas, as well as their policy-relevant actions.

Yet, apart from these activities and challenging duties, I have continued to contribute also to other ongoing dialogues. Sometimes they overlapped with those of the social policy and social welfare community, at other times the connection remained an indirect one. In bringing together a selection of essays written during the years in which the European Centre was my institutional home, I nevertheless can see certain thematic continuities and concerns. They are reflected in the structure of the volume. It has not been possible, either to revise the essays, nor to provide special introductions to the three main sections. For reasons of cost, the selection has also to be limited to articles that I have written in English.[1] Despite these shortcomings, I hope that my readers will find something worthwile in this collection and that they will accept to be part of the utilization context which provides the underlying guiding theme.

Acknowledgements and thanks go to the following publishers for permitting the reprint of articles originally published elsewhere: Longman, North Holland Publishing Company, Pergamon Press, Plenum Press, Sage Publications, Springer Verlag.

The Austrian Ministry of Science and Research is to be thanked for contributing to the publication costs of this volume.

In dedicating the volume to the European Centre and my friends and colleagues during the years I spent there, I wish to express much more than a fare-well gesture. It is with gratitude and humility that I wish to acknowledge what I have learned from them and through the European Centre.

Helga Nowotny

---

1   Thus, more recent work in German on coping with uncertainty connected to social risks as compared to technological ones, had also to be omitted.

Temporal Constraints

From the Future to the Extended Present – Time in Social Systems

Embedded and Symbolic Time

The ways in which societies conceptualize their experience of time is neither immutable nor universally valid. This holds especially for the familiar categories of the past, future and present, categories which are separated by shifting boundaries and different meanings attributed to the guiding orientations they entail. They are not arbitrary constructs, however, since they reflect society's deeply structured experience of time and its collective time consciousness which is echoed in the mind of an individual living in a given historical period and a concrete societal 'Umwelt'. What enables the social construction of time and the expression it finds in both concepts and behaviour, in institutional arrangements as well as in the rich and complex patterns of integration and coordination which characterize a society's temporal existence, are symbolic processes of time structuring which are inter-subjective and constituted through social interaction both on the behavioural and the symbolic level[1]. Institutional mechanisms usually are devised to reinforce them and to work their way through conflicts between different sets of temporal priorities and those patterns of standardization and coordination which form the temporal grids criss-crossing societies. But the powerful capacity of the human mind to form abstractions and concepts of synthesis out of social experience, of reaching beyond immediacy and of being able 'to time' by establishing relationships between two or more moving continua which possess meaning for social interaction[2], is itself restrained. Time is not only a symbolic construct, but it is also, as Torsten Hägerstrand has reminded us again, 'embedded' in the spatial configurations of dead and living matter. Embedded time exists in the unmeasureable succession of more and more complex configurations of matter, in the way in which things appear, transform, change their distribution and disappear[3]. Even our corporality is to be viewed as a part of nature and, therefore, is a carrier of embedded time.

Time is not only embedded in nature, however, but also in artefacts, in the purposeful and instrumental devices that men and women have invented, utilized and expanded over the immense range from early cooking and hunting utensils to the vast socio-technological complexes and mega-systems that determine our technological 'Umwelt' today. It is embedded in the sense that all these artefacts, especially technologies, are not time-neutral or time-independent. Rather, as I will show in a greater detail, technologies constitute an important constraining and determining factor in the ways in which the symbolic processes of time structuring unfold. The construction of social time, itself fascinating because of the powerful play of symbolic interactions and the inventiveness of the human mind, is constrained and directed in specific ways through the time embedded in the artefacts with which we are increasingly structuring and transforming whichever natural 'Umwelt' there once existed.

One of the main themes of my paper will be the exploration of the interaction between the time constraints exercised on and derived from the time dimensions which are contained in technologies and the symbolic processes of time structuring that lead to specific modes of time consciousness. This, however, is not to be interpreted as following the path of technological determinism. Technologies, in the embracing sense in which I will use the term, are an efficient way of using technical means and procedures to alter both the natural and social 'Umwelt' in order to bring about a deliberate change and to achieve pre-set goals. As Marx observed a long time ago, technologies, in accomplishing this function, do not only alter the relationship between humans and the environment, but also the relationship between humans and the things. These changes include the way in which things – artefacts – and other human beings are perceived and experienced, as well as the extent to which they shape interaction and attitudes. But the artefacts are made by humans, they evolve out of specific configurations of skills, knowledge, societal values and economic preconditions, i. e. on the social and economic forces which engender what is called 'innovativeness', and which at present is not yet sufficiently understood.[4]

Control Through Artefacts: The Example of Clock Time

Technological artefacts or the fabricated populations of material artefacts, as Torsten Hägerstrand calls them, have been a highly efficient and ever evolving means for extending the range in which humans could escape certain in-built restrictions. They have served to expand the spatial range which can now be covered by means of transportation and communication just as they have complemented, supplanted and superseded practically all other limitations of our original natural habitus and habitat. They even hold the promise if not of escaping, so of delaying, the inevitable end to our existence, death. Yet, in their real accomplishments as well as in the utopian imagination of their liberating tendencies, they have also continued to exercise new forms of control. They are experienced as constraints which seemingly emanate from the impersonal clock-time which still is the underlying regulator of daily life in industrialized societies. These artefacts are represented by the dark sides of technology, the devastating consequences and oppressive uses they have been put to especially during that intense period of industrial upheaval called the industrial revolution radically transformed the face and fate of this planet. Nowhere has this been better recorded than in the rich historical accounts of the relentless rationalization of time which was introduced when industrial capitalism entered the stage of world history.

The impact of industrialization as a new form of control has been recreated as the history of time measurement and that of the clock – to use Lewis Mumford's well known phrase – as 'the key machine of the modern industrial age'. It brought along a regimentation which, as Mumford also pointed out, was not only essential to the rise of capitalism but also may be considered its product: "the new bourgeoisie is counting house and shop, reduced life to a careful, uninterrupted routine: so long for business: so long for dinner: so long for pleasure – all carefully measured out, as methodical as the sexual intercourse of Tristam Shandy's father, which coincided, symbolically, with the monthly winding of the clock. Timed payments: timed contracts: timed work: timed meals: from this period on nothing was quite free from the stamp of the calendar or the clock. Waste of time became for Protestant preachers, like Richard Baxter, one of the most heinous sins."[5]

I will not dwell longer on the splendid and well-known accounts of the history of Western civilization under the regime of precision time keeping which authors like Mumford, Cipolla, David Landes and E. P. Thompson have provided or extended for purposes of comparison like in Joseph Needham's work. My point here simply is to restate that relationships to artefacts contain important temporal dimensions and produce different temporal norms that regulate and control behaviour. The clock as symbol for the underlying profound transformations which industrial capitalism brought about, its worldwide expansion and its continued rationalization was equated with these processes to the extent that social time became reified in the devices that were used for measuring it. Hated, endured, monopolized and manipulated by different social groups, an instrument of power and expressing the quality of social integration at the same time, it is unsurprisingly often been interpreted in accordance with the logic that is imputed to a reified construct. It has either been analyzed as something secondary to physical time, as the constant process of adaptation of social institutions to an external, objective and pre-supposed reference scale, or as an instrument of oppression. Thus, the critics of the corporate technostructure have equally remained fixed on the clock as the central instrument of domination, especially in working life. While it is certainly true that there exists a policy of time, the clock – like any other piece of technology – neither is a neutral time-keeping device, nor a deterministic force in historical development. Clocks are neither liberators, nor oppressors. The function they serve in our societies is similar to that fulfilled by masks in some tribal societies: everyone knows that these masks are produced by humans, yet in the rituals in which they appear, they are experienced as representatives of a supra-natural entity. Clocks embody time and they show time which is embedded in them as artefacts. Besides, they symbolize the constraints emanating from the production technologies for whose social regulation they have mainly been used.

But how can we come to see the masks unmask, and understand the ritual of which we form an integral part of ourselves?

Past, Future and Present

Time conceptions, as stated in the beginning, vary from one society to another not only because of different cultural and cosmological traditions, but because the different diffusion rate of technologies, the degree and distribution of their societal penetration, the cultural usage and functions they have assumed alter the temporal relationships which govern the interaction between artefacts and humans in a mediated form. This brings me to the second major theme in my paper: the shifting boundaries between past, future and present whereby I will mainly concentrate on the latter two mentioned. Since the onset of industrialization the 'great divide' in the mode of time-structuring which separated industrialized from non-industrialized countries, such shifts can be traced back to the different forms of time discipline and temporal norms that are preconditions for as well as results from the ways in which time technologies impinge upon the social experience of time. Past or future are no longer united in elaborate myths of creation or in cosmological accounts but are contained and shaped to a remarkable degree through the processes in which time embedded in technological artefacts and symbolic time constructions as inter-subjective structuration, interact.

Thus, it is quite remarkable to observe how the 19th century, which bore the brunt of industrialization, was haunted by struggles of coming to terms with both the past and the future. Scientifically speaking, it provided the first firm basis for establishing the facts of nature's very own past: in 1800, scientists knew well that the earth was ancient, but they had no frame for ordering events into an actual history since the primary criterion for unravelling that history, the sequence of unique events forming the complex history of life as recorded by fossils, had not yet been developed. By 1850, history had been ordered in a consistent, worldwide sequence of recognizable and unrepeated events, defined by the ever-changing history of life and recorded by a set of names accepted and used in the same way all the scientific world over[6]. The second half of the century witnessed the triumph of Darwin's 'long argument' in which the past, i. e. history, stands as the coordinating reason for relationship among organisms, as a guide to action in research and the first workable programme ever presented for evolution[7]. By bringing history into science and hence by firmly establishing the order of the past the door was opened wide for a future which was recognized as evolutionary and as open to workable programmes of scientific and technological intervention as well. In the same period, all those individuals who found themselves thrown into the birth pangs of modernization which came with the upheaval of the industrial revolution, realized their previously experienced knowledge of the past to be thoroughly shatterd. It had to rebuilt under scientific guidance in accordance with the norms that came to dominate the world views of a modernizing age. Their views of the future, as Marshall Berman has shown in a moving account based on the cultural experience of the tensions induced by 'le tourbillon social' were tinged by the experience of a relentless destruction of the past, but even in the midst of a wretched present, they could imagine an open future[8]. The experience of innovative self-destruction forced them to give up a past – the innumerable little worlds of small hamlets and towns that survived in an agricultural economy – and to exchange it for a violently uncertain future which they invested with their hopes and fears, their passionate desires and the will to succeed. For the vast majority of those who were torn out of their normal lives, the present was reduced to the most shaky of experiences while the past they had lost proceeded to be reconstructed with the help of science. In turn, this scientifically secured past was used for building a future envisaged as the never ending road towards technological and social progress.

What I find equally remarkable is the fact that the implementation of the future dreamt of in the past, has radically devalued what our own life once held in promises and attractions. Somehow, with many of the technological utopias realized and the miracles in place, the category of the future has also been flattened and exhausted.

The belligerant and overly self-confident claim 'we have seen the future and it works' uttered not so long ago when the idea of planning was advanced on a more massive scale for the first time, has turned out to be a very hallow claim indeed[9]. The present predicaments and threats of nuclear annihilation or environmental collapse, of time running out in multiple instances where – in the wake of technological interventions based on too much ignorance or wilful defiance, probably irreversible processes of environmental degradation have set in – have converted the future into a precarious category, experienced for the first time also in its truly global dimensions. Like similar ventures of this type before, the colonization of the future, so confidently initiated by science and technology, has become recalcitrant turning against its alleged masters. As I will argue in greater detail, it is the discovery and experience of the extended present instead which opens new visions at a turning point in collective time consciousness while we are about to abandon the notion of the future in the sense in which it has been conceived until now.

Time Technologies and Temporal Norms

In the remaining part of the paper I distinguish three types of time technologies. They do not appear in an evolutionary sequence. Rather may they be found co-existing and feeding upon each other with varying inter-connections although major advancements in each of them tend to be concentrated in certain periods. These technologies deal with (1) the reduction of distance in time, i. e. time and space technologies; (2) the conversion of material things in time, i. e. production technologies; and (3) information-intensive technologies which eliminate certain time-space constraints altogether. The dominant temporal norms I see associated with them, are mobility, scarcity and flexibility. By pulling together several strands of empirical evidence and conceptual approaches I hope to illuminate from a social point of view the dialectical processes of timing our minds and minding our time and thereby to enlarge our collective understanding of the societal transition which we find ourselves in right now.

Mobility

Time-space technologies permit the reduction of distance in time, i. e. the transportation of goods, people and energy over geographical distances may be effected in potentially less time. The bridging of space in less time can fulfill the same function as before but it alters its efficiency: armies can march against each other, ride on horse-back or operate automatic missile and anti-missile systems. New social functions – like those that continue to evolve in urban conglomerations – emerge, generating additional infrastructures which have been the object of intensive studies of city-planners, system analysts and geographers. The histories of transportation systems belong to this category with the far reaching spatial and temporal impact on perception and the division of day and night that systems such as the railway or the electrical power grid have had on societies[10]. The possibility of transporting goods over greater distances or in less time has dramatically altered the size of markets, a consequence which usually is not sufficiently appreciated by economists when they speak about market size. The 'timing of space' and the 'spacing of time' have become a powerful conceptual framework for the study of shifting configurations of human activities in geography, pioneered by the work of Torsten Hägerstrand and his collaborators.

The temporal-spatial norm that has emerged from this set of technologies simultaneously incorporating a key societal value and a behavioural prerequisite, is *mobility*. It has come to be highly valued not only in economic, but also in cultural terms. Movement through space in (less) time has become the sine qua non of a system which is based upon the (rapid) circulation of goods, people, energy and information. The pressure to be mobile, to increase the rate of circulation has spread from geographical space to other spaces: mobility in occupational life is inherent in the notion of a 'career'. *Mobility* has become necessary not only in geographical terms for those who want to stay within the labour force when working places themselves become 'mobile' – in being rapidly displaced by new ones or none at all. We widely expect the mobility of ideas to be one of the main springs of scientific productivity. Technical civilization imposes new demands as to being increasingly mobile between different domains, such as work and leisure; high and low culture; the emotional and the rational controlled part of our lives. It is the rapid succession of, or the oscillation between different stages, states (of mind and body) and places of people, things and ideas that has become the dominant time-related norm upholding the rapid circulation of production and consumption of material goods. In addition to being a trait inherent in fashions of various kinds, the appeal of novelties, the increasingly stepped-up succession of what is considered to be 'new', has become the hallmark of material and cultural production alike.

Scarcity

The second major category of technologies with a distinctive temporal dimension is that of production technologies. By differentiating work into separate parts to be performed in equally divisible units of time, they have brought about the incessant division of labour. The amazing gains that have been achieved in productivity since the beginning of industrialization, consist in changes of production methods which allow an increased output from a given volume of labour and resources, or a given output from a smaller volume of labour and resources. From about 1880 onward, the productivity increasing impact of technological advances in the Western economies has been linked by economists to the so-called rationalization of production methods consisting in the rigorous and coherent application of the 'economics of time', i. e. the temporal organization of the industrial production process[11].

The historical record of 'saving time' and the continuous exploitation of time resources within the production process has been meticulously documented as

15

proceeding in twin fashion through technological development and the temporal organization of production, its 'rationalization'[12]. For the present purpose it is important to note how the 'economics of time', or –, in B. Franklin's words, the equation of time with money, spread from their highly successful application in the production process (a fact that was painfully experienced by the workers) to almost any other segment of social life. From the downtrodden workers who protractedly fought for the reduction of working hours to the 'hurried leisure class' (for whom having 'no time' became a status symbol) the impact of the time dimension of production technologies on all facets of life became pervasive: based on the economic value that time acquired under industrial capitalism and the convertibility of time and money, time became the ultimately finite, scarce resource. *Scarcity* of time became the temporal norm that pushed further the efforts towards the opening up of additional temporal resources. In the production process this occurred through further rationalization. By speeding-up the work rhythms of machines – and of the humans that operate them – and by improving synchronization and coordination, additional time was 'gained'. The invention and massive use of the conveyor belt in industrial mass production was but one variant of rationalization methods: by minimizing the time periods of stillstand and rest as well as the slack-time for machines and workers, additional time resources were utilized.

Time as a scarce resource and time discipline, such as punctuality and the general sense of time consciousness have become the hallmarks of the temporal grid of highly industrialized societies. While temporal flexibility and increased autonomy in the use of one's time have remained the privilege of an elite, strict time discipline and the polarization of social life into work and leisure de facto was imposed on the majority of working people.

Further technological advances resulting from the impact of information-intensive technologies are equally causing these patterns to change.

Flexibility

Information-intensive technologies applied to the transmission and processing of information have achieved a new quality of speed, approximating the limits of time that are open to the capacity of the human mind. They are able to perform operations in 'almost no time', i. e. the gain of additional speed is becoming a factor of marginal utility. Concomitantly, information-intensive technologies in many instances no longer necessitate the simultaneous bodily presence of producer and consumer, of sender and receiver. This, for instance, goes for the so-called 'mental services'. This trait has recently been emphasized by Fritz

Scharpf when questioning their potential for rationalization and the opportunities for lowering their production[13]. The customer's wish to have a haircut still necessitates the interaction of hair dresser and customer at the same place and at the same time whereas – given the storage permitting technical means – the taping of concerts, soccer games, or language courses no longer contains such a requirement. Production and consumption can now take place independently from each other. Storage and conservation techniques have already made possible a similar temporal independence for the production and consumption of food: Bio-technology and genetic engineering are proceeding to accomplish this for living cells and eventually perhaps for human life.

The de-coupling of acts of production and consumption which gives way to further rationalization on the side of production, since daily, weekly or seasonal rhythms in consumer preferences need no longer to be taken into account is, however, only one aspect of a more general principle of temporal experience which is embedded in the new technologies. Just as *mobility* has become the primordial temporal experience in the wake of space-time technologies, *scarcity* and the economic value of time, that of production technologies, and *flexibility* will become the temporal norm derived from and buttressing information-intensive technologies. It is also bound to spread to other forms of social organization. Following from the patterns of interaction, work organization, production and consumption links and the like which are imposed by the new technologies, *flexibility* also is a prerequisite for the successful implementation of the latter. Temporal norms are engendered by technologies, yet they become necessary for sustaining the technical system and its social infrastructure once they are in place. Moreover, they are upheld by specific modes of time discipline which, among other things, are set up in order to solve conflicts between different time conceptions and norms.

Flexibility, in a way, obliterates the rigidities of time structures which Western industrialized societies have become adjusted to in the course of the past two centuries. The increase of speed, just like the underlying model of mechanical causal linearity has been mastered and superseded by technological systems that are based on cybernetic principles, emphasizing networks, feed-back loops, decomposition and recombination of component elements which follow the organization principles of complexity, rather than those of mere accumulation and output. These developments and the subtle paradigm shifts that have preceded them within physics, biology and the new cognitive sciences are now becoming more visible in their technological embodiment which also leads to a much more tangible and direct form of interaction with other parts of the social system.

The inherent flexibility of the new technologies manifests itself in an increased availability of possibilities for decentralizing work and consumption in space and time. While this will undoubtedly lead to a continued blurring between work and leisure time witnessed already now in shadow work and tendencies towards 'self-service', it opens up new opportunities for individualized time preferences. However, the euphoria with which the liberating propensies for the new technologies have been greeted as offering an escape from the iron time cage of the past, is premature. Time flexibility, like decentralization, works both ways. Each new level of decentralization and temporal flexibility which is attained in a system, necessitates a new level of greater complexity where coordination, synchronization and centralization takes place. It is, therefore, not to be expected that hierarchies will be abandoned; rather will they become more differentiated and complex. While it remains to be seen how far flexible automation as the factory of the future will spread, there is every likelihood that new work organization principles emphasizing settings in which workers are multi-skilled and paid for what they know, rather than for how hard they work, will bring about a seemingly radical breach with past patterns of work organization and their strict temporal time discipline[14].

The New Time Discipline

Next to work organization there exist culture orientations that support it. According to Donald Lowe, the bourgeois field of perceptions was constituted by the predominance of typographic media, the hierarchy of sensing which emphasized the primacy of sight and the epistemic order of development-in-time which provided a temporal connection for observable phenomena beyond their representability in space[15]. Looking back on it today, it is surprising – to cite but one example – how great an importance was attached to punctualiy by the older educational practices and how drastic the means had to be for archieving it[16]. This occurred in periods in which, judged from today's standards, punctuality was not so much needed. While today, where everything depends on a high level of coordination of individual schedules and their link with the institutionally imposed schedules, such as the traffic system, opening hours of shops, work schedules, etc., punctuality no longer needs to be taught. It has been thoroughly internalized, since the constraints that come from outside and which are imposed by the technical conditions of our lives, are overwhelming.
Flexibility demands a particular state of mind and social forms of organization that facilitate it. The thrust towards 'self-organization' is indicative of one direction that corresponds to the idea of smaller, more decentralized and thus semi-autonomous units. Another direction is indicated by the emphasis that is put on combatting old age. It is highly revealing that this takes place not only with re-

gard to biological age but also in the realm of technological performance. Industrial and political leaders intensively flirt with the process of continuous technological innovation exhorting firms and corporations to fight against senescence. In some cases, like in that of the US car industry, they are promised the possibility of a 're-birth' as an ideal which industries should consciously strive for. Instead of accelerating maturity, American corporations are advised to strategically plan for 'extended product adolescence'[17]. Survival and corporate immortality through management of the life cycle of technological processes and products become new goals.

Finally, there is the campaign Mancur Olson has started against the sclerosis of institutions with similiar metaphors and aims in mind[18].

It will be the task of empirical research to carefully scrutinize the evidence of the internal life time of various technological systems of the kind of Marchetti has pioneered, to determine whether and to what extent, technologies are pre-determined, or – like workers previously – can be "taught" the new time discipline aiming at a prolonged or continuous re-birth of their technical life cycles. Of course, technologies do not behave like humans, they are generated by human action. The anthropomorphic parallels that emerge when studying technological innovations are, however, striking. To me, this suggests that cultural orientations that serve to guide the sense of temporal direction and of time discipline in the broadest sense are no longer addressed to people, but have to include technological artefacts. It is not the one, lonely robot, or Frankenstein's monster that has come to life and that is to be controlled. Rather is it an entire socio-technical system and its interdependencies with society that needs guidance and management. The time discipline which has to be learned, is no longer one of the rigid subordination to clock time or having to adjust to the rhythms of a machine. Rather, will time discipline be built on managed growth and obsolescence, on the planning and control of birth, reiteration and decline. Being part of that system, we are perhaps more consciously than ever before, able to bring along our own biological and social knowledge and experience; we may have a historically unique chance of applying them. Yet it will not be possible either to abstract from the social costs that every time discipline extolls in different ways.

Escape from Contradictions

When Schumpeter equated technological innovation with 'creative destruction', he was mainly concerned with what he thought to be the driving force behind capitalism: the inventor-entrepreneur. This lonely genius has long since given way to the numerous Silicon valleys that have appeared suddenly all over the US, Japan and Western Europe.

Technological innovation has become science-based and science itself can no longer proceed without technological innovation. Yet the vision of what once was conceived as more or less unilinear progress, has become punctuated by Kondratieff cycles, long waves and inherent instabilities. "The gradual, incremental unfolding of the world system in a manner that can be described by surprise-free models, with parameters derived from a combination of time series and cross sectional analyses of the existing system" [19] is no longer considered sufficient for describing, analyzing, anticipating and acting upon the intensifying patterns of interaction between technological innovation, impact on the natural environment and the social fabric. The search for new heuristics of thought, of methods, models and concepts that permit to capture such phenomena as 'surprise', unexpected discrete events, discontinuities in long-term trends, or the sudden emergence of new information in public consciousness and awareness (such as e. g. acid rain) are an attempt to come to terms with the perception of new temporal patterns [20]. They are part of learning to cope with the new time discipline, of having to manage innovation, repetition and decline in their multifaceted forms.

Time in social systems, as I started at the outset, is not independent of 'embedded' time. The one found in artefacts and in their interaction with the human mind. Hence, it is bound to change with the evolution of societies. Just as the heavens – once thought to be inhabited by the gods, where events on earth were but a mirror reflecting celestian happenings – gave way to Newtonian time which has been superseded by the plurality of time in modern physics as described by Prigogine, so is social time undergoing a redefinition. It is structured by the constraints and opportunities that are exerted by the technological systems, but it also has to find solutions for the temporal conflicts that arise from the *co-evalness* of differently experienced needs for coordination and integration. Temporal asymmetries have to be managed through social and institutional mechanisms and through the provision of cultural symbols that serve as means of orientation. In a fine analysis the philosopher von Wright arrives at the metaphorical observation that "time is man's escape from contradiction", meaning that, in order to accept change, time is necessary [21].

Change in social systems arrives with different faces. In the late 20th century, change is dominated by scientific and technological developments, by the fast pace in which nations press for further technological innovation in their military and economic games of competition. A central contradiction exists in wanting to make the unpredictable occur while trying to predict it before or while it is happening. Western industrialized societies have opted for relentless change, yet they want to control its consequences as well as the process through which it occurs. Choosing futures has become a popular theme in very different political circles – its ideological underpinnings range from grassroot political participation to the marriage of market and technocracy where the market creates options of futures and where technocracy is permitted to pick the winner. The answer to these contradictions, the escape route which is opening up, consists in re-arranging our relationship with the future.

The Extended Present

As I see it, we are about to abolish the category of the future as we conceive of it now and to replace it with that of the extended present. We are approaching finiteness in the sense that we have gained sufficient understanding of the problems which beset us in the present in order to stumble towards devising corrective policy measures. The long-term consequences of human interaction with the natural environment has become a paradigmatically challenging case for such understanding, but we can find a sufficient number of examples from any other policy domain once we look for their longer-term interconnectedness. This modicum of understanding and the sense of relative mastery and control due to our scientific heritage closes the future as an inconsequential perspective into which hopes and unresolved contradictions can continue to be projected. Rather, pressure is mounting from many quarters to the effect that problems have to be dealt with now, in the extended present. Just as science in the 19th century was securing a past for nature on which the future of human societies could be built in the late 20th century, scientific endeavours have increasingly to turn towards coping with the problems that beset the present by taking the mortgaged future into account. This, in no way, is contradictory to the fact that the touchstone for mature scientific achievements remains the ability to predict. By extending the present to beyond the immediacy of day-to-day concerns, the range of potential consequences which comes into view, demands, under the pressure of time, that action be taken if negative or irreversible results are to be avoided. This makes heightened demands on science's ability to predict, if only for the simple reason that the possible negative impacts created through scientific and technological interventions, may arrive earlier and possibly may be more acute than before. The quest for the future – to predict and to control it – is catching up with us by loading the present with choices that have to be acted upon under the pressure of an often insecure knowledge and in the face of massive uncertainty[22]. The point at issue no longer simply is in which direction the world will be moved through scientific and technological knowledge, but how the various applications of this knowledge, the uses to which they have been put, are interacting with each other, thereby producing a complexity of results which hitherto has not been consciously experienced in the history of humankind[23]. By successfully reaching into the future, the price for doing so has partly to be paid in the present. Even the line, that separates us from future generations, is becoming thinner. Like any other generation before them, they will inherit a world which is not of their making together with unspecified resources for remaking it according to their own capabilities, desires and conflicts. Yet, the burdens which we are consciously and knowingly inflicting upon them, have become larger and potentially more irreversible. In our still muted concern for their well-being, the future is also intruding into the present and its voices demand some kind of representation now. The scientific task of coping with the problems that are attendant in the present, the boundaries of which have been extended through our dim awareness of the consequences of our scientific and technological doing and the urgency of further remedial action, is further exacerbated through the fact that much of policy- and action-related science has to proceed under novel conditions of uncertainty. In numerous instances, decisions have to be taken with the knowledge basis still insecure and empirical evidence either unavailable or highly controversial and contradictory. The right to not-knowing as yet had been granted without further discussion to science as long as it was seen to advance steadily. Now, under the pressure of public policy decisions which have to be taken with economic and social interests colliding, a new understanding of the precariousness of scientific knowledge in the extended present has also become necessary. The limits to the growth of the present civilization have become internalized and are represented in its knowledge base.

The consequences arising from this diagnosis are manifold. On the technological and social side of the management of technological change, as a matter of consequence, not only innovations, but also decline and obsolescence have to be accomodated in the present. We cannot only plan for change, but plans are equally needed for orderly ways of discarding what has no longer any place in the system. It also follows that long-term time perspectives acquire a definite meaning under such boundary conditions: they become open to operationalization however insufficient they still may seem. The year 2025 becomes a calculable reality on society's global agenda, just like the average manager's desk calendar already includes appointments for the year to come.

Within such an overall time frame, the plurality of times in different segments of the social system from the macro to the microlevel, also takes on added significance. It is no longer so much the individuals' time constants, needs, attitudes and orientations that have to be ordered in a hierarchy of temporal priorities, but the time frames of institutions that need better coordination and a place in a nested hierarchy. Beyond coordination in the same functional level like that of time schedules of traffic flows, work, opening hours of shops, etc., it is the second and third order of time-delayed consequences of institutional interactions that need to be attended.

The abolition of the category of the future and its replacement by an extended present, is not so much of a radical break as it may seem at first. Many societies have lived for long times without paying much attention to this category. The idea of an open-ended future oriented towards constant improvement emerged relatively late in Western history. It implied the creation of an (ideological) distance between what has been experienced and what is expected[24]. With the help of science and technology this distance became stabilized in the 19th century and was identified as the for-ever-evolving dynamic movement of progress. The future became accessible under the condition of remaining inaccessible. Today, this distance is threatening to collapse. Expectations no longer hold the glittering promise of a horizon that is still to be reached and experienced as the basis from which one wants to extrapolate future expectations and thus have lost much of their credence. The category of the future is shrinking towards becoming a mere extension of the present because science and technology have successfully reduced the distance that is needed to accommodate their own products. While science and technology seemingly are producing a multitude of possible futures that are being held out at our disposal, there can still be only one present[25]. It has to be re-structured so as to accomodate the choices heaped upon us through the rapid pace of scientific and technological development. It has to come to terms with the new quality of finiteness which arises from the closure of the future.

What about human freedom under such seemingly restrained conditions? I cannot really enter the venerate philosophical and even theological debate on the freedom of will or the degree of freedom for individual and collective action under the conditions described above without stating at the outset that I find no ground neither for excessive fear nor for technological optimism. As always, new technological developments *or* science putting hitherto unimaginable achievements at society's disposal bring to the fore hidden value conflicts and divergent views on how societies ought to be managed and by whom. The very definition of what is to be controlled and how, opens new space for power. Once again, the struggle as to who will exert it in a society whose innovative contours are still fuzzy, is taking place now. The more the heated discussions and controversies turn around what the case is going to be in the future, the greater is the concern as to who will decide in the present.

In my view, the intense preoccupation with the future as though it were really open for choice, is nothing but the expression of nostalgia for something which is about to disappear and which is thus prized like an antique. Under the pressure of having to cope with incessant innovation – a process which at least in its incipient stages, is governed by randomness – an effort at rationalization sets in which is geared towards more control of the innovative present. The investigation of life cycles of technologies, past and present, enjoys great popularity at the moment[26]. It conveys a deterministic picture of societal life. Yet, Marchetti also alerts to some striking features of self-consistency of structure as it is found on very different levels and in very different segments of the social system. How much freedom this allows, to my thinking will in the end depend on our own definition of where the self in the principle of self-organization is to be located and of what it really stands for.

In the meantime, let me end by approaching what I believe to be the same underlying problem from a slightly different perspective and by phrasing it in more general terms.

Innovation, Repetition, Obsolescence

Reflecting on linked sequences in works of art, George Kubler remarks in an extremely interesting book: "The occurrence of things is governed by our changing attitudes towards the processes of invention, repetition and discard. Without invention there would only be stale routine. Without copying there would never be enough of any man-made things and without waste or discard too many things would outlast their usefulness. Our attitudes towards these processes are themselves in constant change so that we confront the double difficulty of charting changes together with tracing the change in ideas about change"[27]. Although Kubler is concerned with aesthetics rather than with what he calls 'useful' inventions, there are similarities to be found in both. Kubler sees an inherent tension between replication, *i. e.* the desire to return to the known patterns and invention, *i. e.* the desire to escape it by new variation. The individual is sheltered by society against too much variation within an invisible many-layered structure of routine. The other great tension exists between retention and discard. The decision to discard is far from being a simple one. It constitutes a reversal of values since things that once were necessary become litter or scrap. Like in the elaborate tomb furnishings of many ancient cultures, also societies have developed rituals for the retention of things. Today, we are concerned with saving animals, things and knowledge which are about to vanish as part of our civilizing tradition.

Although investigations into the systematic of internal age of artefacts, ideas or technologies which appear in a linked sequence, hold their own fascination, the question of how societies balance their rates of innovation, repetition and obliteration approaches perhaps the same phenomena from a different angle. It is not only a whimsical change of attitudes that tips the balance. With the closure of the category of the future, the present becomes more loaded with all three processes – innovation, repetition and discard. The denser the rate of innovation the more attention will have to be given to the problem of waste and waste disposal. Senescent or dying industries, regions, and especially the knowledge and skills of people working and living in them, cannot so easily be discarded as throw-away bottles. The social problems associated with the loss of social identity, self-esteem, social status and income of those who are made redundant, make the problem of waste appear in yet another form and in the long run thus cannot be left unattended. One strategy of coming to terms with the social costs of obsolescence is to cushion its effects like those of old age. The other strategy consists in trying to eliminate waste on the other end, or rather the beginning of the spectrum among innovations. Rather than welcoming innovative activity unselectively, the management of the extended present demands that a certain amount of order and rationality be put into this process as well. But how is one to know whether the springs of creativity will be weeded out in the right way? Will there remain enough time for time to tell?

Global Finiteness

Time has come to summarize.
● My observations have focused on the interaction of time embedded in artefacts, especially the time dimensions of technologies, and the symbolic processes of time structuring which lead to the establishment of specific temporal norms.
● The newcomers among technologies, information-intensive technologies, have approached the limits of speed that can still be exploited through the old means of social organization of time. By de-coupling previously existing temporal constraints they generate a new temporal norm, that of flexibility. It entails a time discipline that differs significantly from that which controlled the age of industrial production.
● Two basic features of this time discipline are discerned on the macrolevel: the pressure and the necessity of coping in a more predictable and controlled way with the processes of innovation, repetition and obsolescence. Although far-reaching consequences are perhaps most visible and tangible in the case of the management of technological change, they undeniably also exist in the social and environmental sphere. The analysis of empirical patterns of duration ('life cycles') of technologies is part of a growing scientific effort to discern – in order to control – innovation processes that appear to be wasteful under the new time discipline.
● The other feature of the new time discipline is related to the abolition of the category of the future and its replacement by the idea of an extended, but manageable and controllable, present. Under such a long-term time horizon, levelling-off effects, the idea of limits, of surprise, as well as that of choices, time pref-

erences and participation in shaping what nostalgically is still called the future, have their place and are compatible with it. The extended present is beginning to become operationalized.

● Coming to terms with finiteness does not mean closure in the sense of an apocalyptic or millenarian vision. On the contrary, a new potential for temporal variety and for a multiplicity of individual and collective time schemes is being opened up. Its possibilities, including the cultural dimensions of temporal perception and experience, remain to be explored. It is quite likely, however, that we are about to transcend 'bourgeois perception' and temporality and that we enter the global finiteness of the 21st century. In T. S. Eliot's words, written in this century[28]:

> *We shall not cease from exploration*
> *And the end of all our exploring*
> *Will be to arrive where we started*
> *And know the place for the first time.*

Hopefully, it will still be our planet.

References

1   Helga Nowotny, "Time Structuring and Time Measurement: On the Interrelation Between Timekeepers and Social Time", in J. T. Fraser and N. Lawrence (eds.) The Study of Time II, Berlin, Heidelberg, New York: Springer 1975

2   Norbert Elias, Über die Zeit. Frankfurt: Suhrkamp 1984. Helga Nowotny, "Das Machen der Zeit: Zu Norbert Elias' Entwurf einer Zeittheorie". Berlin 1982

3   Torsten Hägerstrand, "Time and Culture", paper prepared for the conference on 'Time Preferences', Wissenschaftszentrum Berlin, Dec. 16–17, 1985

4   Richard R. Nelson and Sidney G. Winter. "In Search of a Useful Theory of Innovation", Research Policy 6 (1977) pp. 36–72; G. Dosi, "Technological Paradigms and Technologies Trajectories: A Suggested Interpretation of the Determinants and Directions of Technical Change", Research Policy 11 (1982), pp. 147–162.

5   Lewis Mumford, Technic and Civilization. New York: Harcourt, 1934, p. 42.

6   See the review by Stephen Jay Gould of Martin J. S. Rudwick's The Great Devonian Controversy: The Shaping of Scientific Knowledge among Gentlemanly Specialists, New York Review of Books, Febr. 27, 1986.

7   Stephen Jay Gould, "Evolution and the Triumph of Homology, or Why History Matters", American Scientist, vol. 7, Jan.–Febr. 1986, 60–69.

8   Marshall Berman, All That Is Solid Melts Into Air. The Experience od Modernity. New York: Simon and Schuster, 1982.

9   Helga Nowotny, "On the Social Ordering of the Future", in: Everett Mendelsohn and Helga Nowotny (eds.) Nineteen Eighty-Four: Science between Utopia and Dystopia. Yearbook in the Sociology of the Sciences. Dordrecht: Reidel 1984.

10 Wolfgang Schivelsbusch, Die Geschichte der Eisenbahnreise: Industrialisierung von Raum und Zeit. Frankfurt: Suhrkamp 1979; Lichtblicke, Frankfurt: Suhrkamp 1984.
    Thomas H. Hughes, Networks of Power. Baltimore: Johns Hopkin University Press 1983.

11 Nathan Rosenberg, Perspectives on Technology. Cambridge University Press, 1976. Sam Macey was right, however, in pointing out that an established clock industry flourished well before this date. Personal Communication.

12 Among the more recent overviews, see Werner Bergmann, "Das Problem der Zeit in der Soziologie",

Kölner Zeitschrift für Soziologie und Sozial-Psychologie", Jg. 35, 1983, pp. 462–504; and especially Jürgen Rinderspacher, Zeit, Arbeit und Belastung, Frankfurt: Campus 1985.

13 Fritz Scharpf, "Strukturen der post-industriellen Gesellschaft", Arbeit und Gesellschaft, 11. Jg. Heft 1, 1985, pp. 9–34.

14 Larry Hirschhorn, Beyond Mechanization: Work and Technology in a Postindustrial Age. MIT Press, 1984; Robert U. Ayres, The Next Industrial Revolution, Cambridge: Ballinger, 1984.

15 Donald M. Lowe, History of Bourgeois Perception. Chicago: University of Chicago Press, 1982.

16 Gernot Böhme. Coping with Science. Towards a New Theory of Science. University of Linköping: Theme T Report 9, 1985.

17 Robert Ayres, op. cit. p. 243.

18 Mancur Olson, The Rise and Falls of Nations, New Haven, Conn.: Yale University Press, 1982.

19 Harvey Brooks, "The Typology of Surprises in Technology, Institutions, and Development" in W. C. Clark and R. E. Munn (eds.) Sustainable Development of the Biosphere. Cambridge: Cambridge University Press, 1986.

20 W. C. Clark, in: Clark and Munn, op. cit.

21 C. H. von Wright, Time, Chance and Contradiction. Cambridge University Press, 1968.

22 This theme is further pursued in Adalbert Evers and Helga Nowotny, 'Über den Umgang mit Unsicherheit'. Frankfurt a. M.: Suhrkamp, 1988.

23 Torsten Hägerstrand, "Presence and Absence: A Look at Conceptual Choices and Bodily Necessities", regional Studies, vol 18,5, pp. 373–380.

24 Niklas Luhmann, Gesellschaftsstruktur und Semantik, Frankfurt a. M.: Suhrkamp 1980; Reinhart Koselleck, Vergangene Zukunft, Frankfurt: Suhrkamp 1979.

25 Helga Nowotny, op. cit. "On the Social Ordering of the Future", in Everett Mendelsohn and Helga Nowotny (eds.) Nineteen Eighty- Four: Science between Utopia and Dystopia. Yearbook in the Sociology of the Sciences. Dordrecht: Reidel, 1984.

26 Cesare Marchetti, "Society as a Learning System: Discovery, Invention and Innovation Cycles Revisited", Technological Forecasting and Social Change, vol. 18 (1980), pp. 267–282; R. Nelson & S. G. Winter, "In Search of a Useful Theory of Innovation", Research Policy 6 (1977) pp. 36–76; Burton Klein, Dynamic Economies, Cambridge: Harvard University Press, 1977.

27 George Kubler, The Shape of Time. New Haven and London: Yale University Press, 1962.

28 T. S. Eliot, Four Quartets. London: Faber & Faber, 1944, p. 59.

The Public and Private Uses of Time

In the past decade, a great upsurge of everyday life concerns has manifested it-self in many ways. The private arrangements and networks that constitute the pre-contractual conditions of the social contract and that are indispensable for the functioning of the public sphere, have come to the fore, carried by new voices that demand to be heard. These are voices raised at the local level where the pressure to continue with an ongoing process of democratization has been felt most: it is here, within relatively small political and administrative units, that de-mands for greater participation in the formulation and implementation of poli-cies, which touch the lives of citizens directly and immediately, have found their most tangible expression. These are the voices of lay people, the non-profes-sionals who ask for accountability of the professionals and for having their say in matters that concern their bodies, their state of health and the extent to which they can and should organize themselves in mutual self-help. Interspersed with, and yet distinct from these various movements are women's voices who have increasingly entered the market of paid labour and the public sphere while con-tinuing to carry more than their share of the load of responsibilities for the pri-vate well-being of their families and themselves. They have brought with them to public life experiences of their own as well as life styles that differ distinctly from those of men and the model of man which tacitly served as norm for many social policies and their distorted claim to universalism. Yet the process, through which these voices and the claims they raise can be accommodated into the existing structure and policies, has been an uneasy one.

Nowhere has the necessity and even the political wisdom to incorporate the new social actors and their potential contributions into the contents of existing politi-cal programmes and administrative practice been so obvious as in the fields of social policy and of the social services in particular. In economic policy the in-formal sector with its 'shadow work' and its 'hidden economy' rarely speaks for itself; the shadow workers, so to speak, usually remain silent; they are objects of analysis and perhaps of official policies, but do not seek to become acknow-ledged partners. In the social field, however, community action groups and local project initiators, self-help organizations and issue-oriented initiatives seeking to reshape welfare from below, have flourished in the past years. In no European country has the welfare state ever held a monopoly on providing welfare ser-vices and in organizing systems of care. It has, however, often remained silent

Reprinted with permission from Balbo L. and Nowotny H. (eds.), Time to Care in Tomorrow's Welfare Sys-tems: The Nordic Experience and the Italian Case, European Centre for Social Welfare Training and Research, Vienna 1986.

on these issues leaving it to local authorities, regional administrative bodies and voluntary agencies of different origins and political persuasions to organize care and to fill the gaps left by state bureaucratic arrangements. With regard to what the clients themselves – the largest part of them being women – have contributed to the effective functioning of these services in the field of health, education of children, care for the elderly and many other highly diverse forms of caring, the silence of the welfare state has been most profound. As Laura Balbo reminds us in this volume, clients also have contributed to and not only taken from the welfare state: irrespective of whether or not – and this is debatable – a culture of caring has evolved in this unacknowledged interaction.

Historically speaking, the silence of the welfare state on the informal aspects of care and social services is not surprising at all: the emergence of the welfare state was dependent on the existence of a national state anxious and willing to add functions to its growing administration. Resulting from a complementary process in the evolution of a relatively coherent, tripartite structure consisting of the state, a capitalistic market economy and democratic mass polity, the state assumed the obligation – as counterpart to the loyalty expected from the institutions of democratic mass polity – to ensure the basic social protection through a system of monetarized benefits, constructed in a subsidiary manner. Thus, social policy remained tightly linked to economic policy. The logic of capitalistic accumulation and expansion may be considered both the cause and the continuing precondition for the existence of the welfare state.[1] As a matter of fact, and as an ideological tenet which sustains it, only economic growth was considered a realistic and viable guarantee for the distribution of welfare goods thus making capitalistic accumulation a precondition for its legitimacy. It is no coincidence, therefore, that the first signs of the alleged crisis of the welfare state, interpreted as a financial crisis, followed immediately after the oil shock had signalled the onset of an economic recession in Western industrialized nations and that this period coincides with the discovery of the informal sector.

Coming to terms with the illusionary promise of continuous economic growth has meant shifts in the 'welfare mix', i. e. in the characteristic combination of resources from the market and the state, from public and private sectors of welfare in order to cover the whole gamut of 'welfare products' as well as the search for different modes of distributing financial burdens and the reorganization along the vertical axis from the local level upwards or the national level downwards. These shifts take place in practice, unplanned for the largest part and often under the immediate pressure of budgetary cuts or changes in political programmes. So far, however, no clear-cut picture of the overall direction amidst considerable national variations has emerged as yet.[2] Nor has the acceptance of innovatory social initiatives on the part of the established social services been an easy or even process: while groping for criteria that would permit the evaluation of these initiatives in terms fitting into the administrative structures and the political background in which they operate, structural deficiencies of a far-reaching nature are becoming more and more evident. It is in this connection that a study like "Time to Care" has acquired relevance.[3] Based on systematic reasoning and being avowedly normative in intention, it spells out constraints and options that will impinge on the welfare state of tomorrow. Since demographic changes entailing a growing number of elderly and the trend towards more women working outside the home, thus reducing the pool of unpaid care providers in the home, are not unique for Sweden, these are indeed developments which will have to be faced in somewhat different forms and with different timing by practically all welfare systems.

In search of better utilization of human resources one is bound to stumble upon that ultimate finite human resource: time. In the official discourse, dominated by the economics of time which has turned time into a scarce commodity, a purely economistic duality has prevailed: the distinction between work and leisure time. Based upon the underlying model of production and reproduction, and later that of production and consumption, it truncates life into two periods: time spent in working for pay (production) and the necessary time to spend in resting (reproduction), or in consuming free time – leisure. While the historical origins for this conception of time are well known, its appropriateness in grasping a changing reality has increasingly become blurred.

Let me elaborate. Neither the production-reproduction model, nor the distinction between working time and leisure time makes room for time spent on neither. These are the periods or retirement in old age, the time women spend on raising their children or the time spent by the young generation for obtaining an education, undergoing training or the time they spend while waiting for their chance to enter the labour market. Increasingly, there is also the unvoluntary unproductive use of time of the unemployed, both in an economic and social sense. Although the time mothers spend with their children may perhaps be interpreted as an economic investment into their future, such a notion captures only the least important aspect of having and raising children today. The time of the elderly can be said to be pre-paid time since old age is financed through earlier contributions of those who have worked gainfully employed or who benefit from a universally financed pension scheme. But the elderly do not only indulge in leisure nor in consumption nor are they overly socially productive; rather are they busy in maintaining their health and well-being. Undoubtedly, the unemployed experience the heaviest losses though they have plenty of time at their disposal. As Marie Jahoda has shown so clearly, they do not only lose a job but also the social structure that is connected to it – social relations and communication with others and the social structuring of time that regulates our daily lives.[4] As seen over the life span, the majority of citizens in the highly industrialized countries spend indeed a good part of their life and time working for pay. Life time is exchanged against money for working time. Since employment typically occurs outside the home, it has become associated with the public sphere. Yet all of us also spend a considerable part of life and time in growing up and growing old and both periods tend to become longer. Since these periods in our lives tend to concentrate on the home, although a good part of them is lived in institutions as well, they have not been granted the status of public time; rather are they subsumed under its private uses. However, interspersed with such a life span conception, they are highly unevenly distributed oscillations between public, paid, and private, unpaid time. Typically, mothers do not only spend more time with their children and on doing the housework than fathers do. In addition, they have to do so in patterns dictated by the needs of their children and their time schedules. Hence, their private use of time is much less flexible than that of a high income earning male who, even if he works a lot, enjoys the maximum freedom as to the disposal of the free, private time at his command.

The point is that there exists a highly uneven distribution of uses of time throughout society and a highly uneven system of exchange mechanisms between public time that is usually spent at work and being paid for and the private uses of time. While it is obvious that some publicly paid time is the precondition for the private uses of time, a good part of the private uses also form the precondition for sustaining the public ones. Laura Balbo has called attention to the "crazy quilts" that women fabricate through the informal activities and uses of time spent in the home, in schools, in hospitals, in the offices of welfare agencies and other bureaucracies – coming and going, waiting, filling in forms, picking up children, etc. While money can be used to buy time, there also remains time that is free in the sense of not being subject to the monetary nexus. There exists an exchange system of this free time which obeys to exchange mechanisms other than those of the labour market: to norms of reciprocity among family members, friends or neighbours, to norms of solidarity in organizations and associations and to common social projects in which people are engaged on the basis of their convictions, beliefs and sympathies.

While the transfers that exist between publicly paid time and the private time at the disposal of the individual remain at the core of economic and social policies, to take 'time-off' working usually has to be 'paid for' by foregoing additional income provided that there exist sufficient work opportunities. But the time thus 'saved' has an economic dimension as well as a social one.

In economic terms, the exchange unit 'time' remains couched in monetary terms. What the individual does with her free time, is interpreted as consumer's choice. Yet private time and its uses are subject also to other, non-monetary norms and forms of exchange. "Care" writes Marten Lagergren, "to a large extent is a matter of time devoted by one person to another." And – one is tempted to add – this is likely to remain so for quite some time. The rationalization process which has proved so highly effective in the production of goods and which has penetrated, though at a slower rate, the service sector of the economy as well, encounters the resistance of the informal services and of care in particular. Despite the inroads the new technologies have made already or are going to make into the home, community and family life, smiles cannot as yet be exchanged via computer, nor can the warmth of a hand, the non-verbal support in interpersonal relations be easily portioned, stored or dispensed at fixed intervals. It can be organized, though, to a certain extent. Voluntary and professional organizations alike have attempted to include at least some of these elements into the canons and practice of both professionalism and organized voluntarism. In the end, however, the crucial question returns: if care is to a large extent a matter of time devoted by one person to another – whose time is it? Is it publicly paid time or private – unpaid and non-monetarized – time? Is it a professional's highly paid and also precious time in terms of demands that are being made upon her, or is it the luxuriously free time of family members or friends, neighbours, of lay people who have come together to spend time with and for each other on the basis of mutual exchange? Is it time which has a monetary tag attached to it, however, discretely it may be hidden, or is it time which does not equate money either because we have decided to separate it from the monetary nexus, or because it is time which cannot be paid for?

Finally, what is the social basis for devoting time to others? Which rights or obligations can others claim on us or we on them? Parents used to invest in having children (not always through devoting much time to them) because they reasonably expected to be taken care of in their old age, at least in the days before the social security system was universalized in industrialized countries. Children responded to these expectations since they in turn expected to be taken care of when becoming old. But what will become of this and other expectations when the underlying standard family model no longer corresponds to a changing reality which has already happened?

In historical terms, the welfare state has been highly successful in providing universal access to a wide range of services in the fields of health, education and welfare. But, as many of its present predicaments show, nothing can fail like success. Compared to the harsh, conflict-ridden realities of its formation, the welfare state has also become something of a myth. In societies where it has been accepted and supported by strong political consensus, as has been the case for the Scandinavian model, it has become a myth which has released quite a bit of highly productive social energy. In other countries the myth has served as a convenient target for attack – the alleged 'welfarism' has become a favourite shibboleth in certain places. In yet other contexts, the emergence of the welfare state has been delayed, the welfare state later followed its own route, committing itself neither to the strong strand of egalitarianisms of the Nordic models, nor following a sharp division between state and market, but becoming enmeshed with already existing social structures and institutions.

While the most public part of the welfare state – the systems of social security and monetary transfers – essentially rests on the abstract notion of solidarity through taxation which is but a form of – unequal – redistribution of publicly paid time, the informal side of the welfare state has far too long been taken for granted, especially the continuing existence of the old models of solidarity to be practised among family members. As several of the papers presented in this volume point out, it is partly due to the successes of the welfare state that the old forms of solidarity in the informal sector are crumbling. These successes have been marked among others by providing higher levels of education and changes of life styles by offering more paid positions in the service sector, especially in the health field, in education and the social services – positions that have largely been occupied by women continuing to seek paid employment outside the home. The family has indeed "gone public" in terms of relinquishing a large part of its previous functions in education, health care and care for the elderly. This has meant universal access to these services, at least in principle; the emergence of professionalization and, again at least in principle, high quality standards of professional care. It has also meant bureaucratization and a functional division of tasks as well as of authority between the providers of these services and the receivers, the clients. Publicly paid time took precedence over private time which was considered gratuitous and of less value. Solidarity and reciprocity in the informal sector remained a residual, almost invisible category. The shock came with the realization that publicly paid time cannot completely supplant the private time of the individual and its devotion to the care of others, nor is this likely ever to be the case. The value of private time is since then on the rise following the laws of scarcity. It appears as an almost untapped resource, especially when connected to the reduction of paid working time calculated both in hours per week and over the entire life span. But is this hope justified? Why should anyone devote time to the care of others rather than to herself or to other, non-caring activities?

There are moral arguments which can be advanced in answering this question but also pragmatic ones and, last but not least, political ones. The concept of a 'culture of caring' attempts to cut through all these categories. Whether it is only a concept, or even a myth, or has a basis in the reality of every day life, remains to be seen. But even myths, as we have seen, can be productive in unclenching social energies. The Scandinavian experience, at least as it is reflected in the papers presented in this volume, points towards a more cautious, pragmatic reality, exemplified in notions like 'caring contracts' or life cycle policies. Helga Hernes points our that privatization, a glittering hybrid concept in itself, can only be one of the many strategies to deal with the problems of administrative and fiscal overload since the interests of those who need care and those who give it – whether they are paid or not – are by no means identical. But can they be made more complementary? Does the proposal to tax the citizens' time instead of their money, carry any chance of being realized and what would its unanticipated effects be like? Would a proposal for an exchange market for giving care and receiving care where time is 'paid in' and 'taken out', much like in an insurance system, fare any better?

I have started this introductory essay with the observation that in the recent past, everyday life concerns have surfaced to a new collective awareness. One of the hidden tenets in the pre-contractual arrangements of the social contract is undoubtedly that we, as social beings and citizens, are not only the victims of time constraints imposed upon us, but that we also take an active part in shaping and structuring time. More complex societies impose greater demands in terms of coordination and integration of individual time frames and schedules. More time is spent in commuting, for instance and on adapting one's needs to the official opening and closing hours of institutions. Besides, it is necessary to spend more time on coordinating the growing number of activities and their greater temporal density. Yet, I would also argue that our faculty of commanding the time at our disposal becomes potentially greater. This goes also for the scope and degree of freedom with which we can structure our time, the range of strategies and competence in coping with temporal and spatial constraints.[5] This has an individual and a collective dimension: individuals – depending on life circumstances, economic means, and on whether we are women or men – still have a far too unequal command of the time at their disposal. Collectively, there are political and cultural decisions to be taken. Following conflicts and negotiations and eventually reaching some kind of consensus, the outcome will be highly decisive for determining the quality of time that will be devoted to care in tomorrow's welfare systems.

However, we should not forget one of these days the other may turn out to be me.

References

1 Nowotny, H. (1985), The Welfare State Revisited, Development: Seeds of Change, 2, 44–48.
2 These and other questions are being dealt with in the research project Ideology and Practice of the New Welfare State of the European Centre for Social Welfare Training and Research, Vienna.
3 Time to Care (1984), A Report prepared for the Swedish Secretariat for Futures Studies, M. Lagergren (project leader) et al., Pergamon, London.
4 Jahoda, M. (1982), Employment and Unemployment, Cambridge University Press.

Time Structuring and Time Measurement: On the Interrelation Between Timekeepers and Social Time

Introduction

At first sight the interrelation between the two main themes of this paper, time structuring and time measurement, seems to be simple enough. Time is something that we measure and that we measure *with*. But what is it what we measure and how is it constructed that we come to think of it as being measureable? As Leach has pointed out, in any society the prevailing ideas about the nature of time and space are closely linked up with the kinds of measuring scales which are thought to be appropriate. If we alter the scales and dimensions with which we measure, we seem to alter the nature of that which is being measured, as well.

Our scientific culture makes us believe that we stand on firm, objective ground insofar as we know what it is what we measure. The prevailing conception of time is largely one of a unitary time, which can be broken down into sub-units of equal, measureable duration. If we follow Leach and wish to draw conclusions from the nature of the scales that we use to measure time with, to our prevailing conceptions about time, then it appears to be significant that we operate today with more precise scales, but with far fewer dimensions than ever before. The history of timekeepers, as told in the prior papers, suggests that this has not always been the case. In the absence of a common denominator or medium in which all values can be expressed in terms of interchangeable numbers, men have operated with many time dimensions and with different scales of measurement.

The history of time-keeping, however suggests something more: in their time-keeping operations, men have always used a model, a hypothetical, external reference scale upon which they could map their conception of what time is. Examples are the movements of heavenly bodies, the natural cycles of agriculture, and the presumed activities of the gods. Today, as befits an age dominated by scientific thinking, precision and unambiguity, we have agreed to the definition of a standard second in terms of a spectroscopic frequency. The impact of

Reprinted with permission from Fraser J. T. and Lawrence N. (eds.), The Study of Time II, Springer-Verlag, Berlin, Heidelberg, New York 1975.

this uncritical quest for physical precision is evident in many facets of our daily lives.

The preference for the physical-scientific methods of time-keeping notwithstanding, there remains current among users of clocks an intuitive feeling about the measurement of time based on personal and social experience. From a sociological point of view, the physical, 'external' time, to which industrialized society tends to refer, becomes a social construct whose nature is somewhat arbitrary and which can be shown to vary in accordance with certain other, dominant features of the general socio-economic conditions of a society. We shall call this inter-subjective, common, time experience which is shared by individuals living under similar social circumstances, social time. Apart from the subjective psychological experiences, grounded perhaps on some information processing mechanisms common to all human organisms (Michin, Toda, Ornstein), there exists however an inter-subjective, social experience of time. We can even go one step further: social time comes into existence through the processes of social interaction on the behavioural as well as on the symbolic level (Sorokin & Merton, W. Moore, Munn). In its most obvious manifestation, the social component becomes evident through the need to coordinate social behaviour and interaction. Indeed, many of our actions are not meaningful in terms of time except when other people are involved. But human interaction is not carried out on the behavioural level alone. What is additionally involved and interwoven into the behavioural fabric are processes of symbolic interaction. In the history of social time, these processes of symbolic interaction can be exemplified by the different conceptions of inter-generational time cycles, and by norms regulating interpersonal temporal behaviour, such as 'moving ahead of each other' (Munn). Not only are relations among individuals, but also relations between individuals and their gods and other idols are involved. Here the link to timekeepers is quite obvious. Very early clocks were probably not used primarily to keep time to facilitate and coordinate social activities, but served as devices made to reflect the orderly sequence of cosmic events, partially reflected by happenings on earth (de Solla Price).

The functional components of social time, born out of necessity and developed in accordance with the needs of modern society, appear to be obvious. But the symbolic processes of time structuring, which are more intriguing, more subtle, and far more consequential in their social manifestations are unobvious. We wish to explore, therefore, some of the mechanisms through which social time is structured. The term 'social time' refers to the experience of inter-subjective time created through social interaction, both on the behavioural and symbolic plane. In accordance with its communal nature, different societies and groups within a society develop specific, variable forms of social time. The underlying social process through which the forms of social times are generated are referred to by the literature as 'time structuring'. It is of interest to note that certain distinctions, identified by psychologists as operating on the level of individual consciousness, appear to have social correlates, mouided by social factors. Thus, for instance, past, present and future may be found on the individual as well as on the social level but with differing significance. In their social forms, the drawing of boundaries between these segments of time, and the meaning attributed to them, as well as the relative weight they are accorded in general, can be shown to be the outcome of social conditions and processes. Societies, or groups within societies, can be found that are primarily past- or future-oriented, as well be shown below. Short-term and long-term time perspectives or time horizons are another example of time categories which appear, but differ in their individual and social significance. There seems to exist a physiological basis for a distinction between short-term and long-term memory (Michon, Deffenbacher & Brown). But whether a society, or a group within it, possesses primarily a short-term or long-term time perspective is a socially and not a physiologically determined fact.

In the context of the present paper we shall content ourselves with an examination of some of the underlying social mechanisms illustrated by these two examples: how do societies, and groups within them, structure social time with regard to (a) the distinction between short-term and long-term time perspective, and (b) the distinction and relative weight accorded to past, present and future. After a brief critical review of the literature, we shall turn to a discussion of some fundamental aspects of time structuring as exemplified by these distinctions and will return later to the interrelation between time structuring and time measurement.

The Poor Have Less Time: On Time Horizons and Time Orientations

An impressive number of research findings on the study of different time horizons and time orientations, especially among different groups within a society, could be summarized in these words: the poor have less time; they tend to live in the present. Other findings suggest two other simple conclusions: industrialized societies tend to be future-orientated; non-industrialized ones are past-orientated.[1]

In one of the earliest studies on time and society, LeShan has given the following description of differences in time horizons among different social classes in the United States:

"In the lower class, the orientation is one of quick sequences of tension and relief. One does not frustrate oneself for long periods or plan action with goals far in the future. The future generally is an indefinite, vague, diffuse region and its rewards and punishments are too uncertain to have such motivating value . . . In the upper-lower, middle and lower-upper classes, the orientation is one of much longer tension-relief sequences. As the individual grows older, he plans further and further into the future and acts on these plans . . . In the upper-upper class, the individual sees himself as part of a sequence of several or more generations, and the orientation is backward to the past . . ." (LeShan 1952: 589)

These findings, by and large, have been confirmed by others, as may be seen in the thorough survey by Moenks. The differences which have been found seem to run straight through various areas of cognitive performance. Fraisse quotes LeShan's report that middle-class children who have been asked to invent stories tell stories that cover a longer period than those invented by working-class children. Bernstein has demonstrated that the time span of anticipation of the working-class child differs from that of a middle-class child. The differences are already embedded in the sophisticated ways which linguistic codes are employed by each social class. Barndt and Johnson have shown that delinquent boys have shorter future time perspectives than a control group of non-delinquents. Nuttin, in a general attempt to link differences in future perspectives to differences in sex and age, has proposed as a working hypothesis "that the depth of the future time perspective in human motivation is not primarily related to age and to differences in age as such, but rather to the nature of the behavioural plans and tasks and to the social structure in which these plans and tasks are embedded." (1973: 72)

It must be pointed out with Rammstedt, however, that many studies on time differentials among different social groups, and on deferred gratification, can be criticized from a sociological point of view, not only on the basis of their often inadequate and superficial methodology (a criticism which cannot be pursued further here), but also on the ground that their authors often became the victims of their own position in the social structure. This finds expression in their frequent assumption that a long-term perspective, until recently found mainly in industrialized countries in a middle-class milieu, constitutes the 'normal' socio-psychological standard against which 'deviations' can be measured. In the literature on deferred gratification (e. g. Schneider & Lysgaard, Singer, Wilensky & Mc.Graven), it is too often assumed that there exists a unitary future, one which is equally valid for all social classes and totally unrelated to the specific socio-economic and political situation in which the individual finds himself. It has been shown by more sensitive social scientists (e. g. Lewis; Miller, Riesman & Seagull; Ortiz) that a short-term time perspective, such as is held by certain cultural minorities, the working class and some segments of the peasantry is far from exhibiting the 'irrationality' often imputed to them by middle-class social scientists. On the contrary, it often constitutes the only rational strategy to survive in an environment which is to a high degree uncertain, loaded with risks that are beyond the individual's control and influence, and about which only a minimum of information is available. Where the future prospects are dim and the very idea of future induces nothing but a state of anxiety, as was the case for a large proportion of the working class in the 19th century (Rezsöhazy 1957) – what need or reward is there to lead one to an expansion of the temporal horizon?

Yet, a reference to the objective conditions alone does not constitute adequate explanation for the phenomenon of different time structures. Socio-economic conditions, like any other 'objective' feature of the environment, are mediated through cognitive structures and representations as well as through symbolic processes of interaction and may lead to the creation of what appear to be exceptions. Sufficient empirical evidence exists to show that neither working-class individuals nor members of cultural minorities lack the ability to transcend the immediate present with its needs and frustrations. Among the many counter-indicators, we find, for instance, that elaborate provisions for specific events are sometimes undertaken by members of these groups. Such events may be the individual's own funeral or a daughter's marriage, which are often connected with great expenditures; such preparations can bring forth attitudes towards savings that rival those extolled by the Protestant Ethic (Holmberg). Another example would be that of revolutionary movements which can take on the character of millenarian movements, especially for the lower strata of a society (Talmon, Worsley). This again suggests that members of the lower social strata are capable of developing long range time perspectives.

Nor can it be maintained that traditional, agricultural societies show no concern with their future, although this concern is likely to take on a form which differs from the one familiar to an industrialized society. Bourdieu (1963), in a study of North African peasants, has asked: how is it possible that the inhibition against scrutinizing the future sanctioned by custom and religion alike is reconciled in practice with such economic necessities as making provisions for a bad harvest or making decisions which inevitably lead to activities located in the future. Bourdieu's research leads him to the conclusion that economic decisions and planning activities among the peasants of Kabylia are based on traditional models of behaviour which have been transmitted from the past and which serve to organize the future as well. It cannot be said, therefore, that future orientation does not exist in this type of society, although it is likely to take on a different form and is closely modelled after experiences that have been made in the past.

In a similiar vein, evidence accumulated by anthropologists suggests that the time spans covered by belief systems in traditional societies are often rather extended ones, even if predominantly they are oriented towards the past. But the function of this orientation is also often tied to the present. Genealogies, for instance, dating back several generations, are minutely organized and cared for, but they have been shown to serve not primarily as a store for oral history, or as a backbone for societal identity, but mainly as a basis for legitimizing the claims to power of those who live now (Goody & Watt).

The brief excursion into the literature warns us against jumping too rapidly to unwarranted conclusions. The evidence does not permit any simple causal explanation, such as the reduction of differences in time perspectives and orientation directly to socio-economic conditions. It confirms, however, that such differences exist, both with regard to time orientation of different types of societies (notably between industrialized and non-industrialized ones) and time horizons of different groups within a society. These, however, are not given once and for all, but tend to vary. In the following we shall attempt to provide a more adequate explanation for these variations by raising first some fundamental issues regarding the mechanisms through which social time is structured.

The Value of Time: Abundance and Scarcity

The most fundamental dividing line between modern, industrialized societies, and traditional, non-industrialized ones, appears to lie in the *value* accorded to time as such. Under certain conditions, time may be judged as valuable or as having no intrinsic value, as being relatively scarce or abundant. Comparative studies on the uses of time, especially as carried out with the help of time-budget studies (Szalai), have brought forth evidence that shows that in economically less developed countries time is not being accorded the same value as in highly developed countries. Research carried out on the Ivory Coast by Vercauteren illustrates this very well. In Korhogo, an urban centre with predominantly commercial and administrative functions, the merchants are well aware of the value of money, but not of time. They display a considerable amount of commercial spirit and employ astute means to attract clients. Thus, they do understand the causal linkage between persuasion and earnings. However, no evident link may be found between the way they organize their activities in terms of time, and the amount of money they can earn. This is in sharp contrast with comparable attitudes found in industrialized countries. This work appears to constitute for them still a multi-functional activity not (yet) geared exclusively towards productivity.

This and similar evidence (Szalai) suggest a link between the value accorded to time and the experience of its relative scarcity and abundance. In a traditional society, time does not possess much value as such. The members of a traditional society can be said to 'possess' plenty of time, especially as compared with the harassed individuals of an industrialized society. This observation is supported by studies which have focused on certain 'losses' or 'gains' in value of time. In a classical study on the effects of unemployment (Lazarsfeld et al.) it was shown how all of a sudden time lost is value for the unemployed, although they had no plenty of it – or maybe because of this. In a study on the effects of prolonged hospital treatment, Roth showed how time gained suddenly in value and how 'time out' acquired new qualities. It became a sort of prized commodity, for which patients would start to bargain with their doctors.

We are thus led to the tentative conclusion that time will be accorded value whenever it comes to be considered rare from the point of view of the individual. But this is likely to remain a purely subjective experience, as long as the meaning attributed to time is not shared by other members of the society or group. Temporal experiences, as Bourdieu (1972) has pointed out, have to be brought together developing correlations between temporal symbols and other mediating devices before they can be used in subtle and highly skilled ways as strategies to build up social relationships. To the extent that these are temporal experiences taken as given, and are regarded as external constraints, they may acquire the character of something sacred as Durkheim has shown. The sacred, however, cannot be meddled with, or be acquired by humans. As long as the society is the "keeper of time" through its macro-level institutions and its "high priests", the question of the individual member according value to time does not even arise. All that human activities can do then is to conform to these sacred, societal patterns and to fit themselves into the "temporal niches" that have been provided for them. In most societies strong social norms can be found which regulate the use of time when and how it is to be filled and with what kind of activities. Our present day calendars are relics of what used to be elaborate societal time-tables, telling the individual what the right moments were for the performance of certain acts, to be embedded in the constant flow of sacred, macro-societal time. Many elaborate precautions and efforts were undertaken, in order to find out precisely what these right moments were – be it in order to wage, to marry or to beget children.

Needham, Ling and Price show the extensive efforts made by the Chinese ruling dynasty in the 11th century, designed to identify the right moment for proper succession. The Chinese emperor not only possessed the most sophisticated clocks available at the time, but he lived according to a schedule putting forth the order in which he was to spend the 365 nights of the year with his 121 wives. On more mundane levels, the same concern reappeared in many different forms, all directed to find out how human activities could best be woven into predetermined, temporal patterns or schedules of societal time. As long as society is the keeper of time, as time was bound up with the whole life of a society, there can be no question of time's being valuable or valueless for an individual, for the determination of the value of time is beyond his control.

We have noted above that the value accorded to time separates industrialized from non-industrialized societies. We have seen now that in societies in which a conception of sacred, societal time dominates the question of an indiviual according value or no value to time, does not pose itself. Rather, the relation of individual to societal time is one of following the right order, of finding the temporal niche into which one could place the activities over which individual control could be exerted. But how, then, did change come about, leading to the phenomenon of regarding time as a prized commodity, one which is on the verge of replacing money?

Time and Economics: The Extension of the Social Present

There is much convincing evidence to support the hypothesis that time acquired its value, in the sense in which this is understood today, through that change in the history of economic development, when time was discovered as a factor in productivity. It was the moment when time came to be used as a medium in which more of something could be produced (Rezsöhazy 1970). The secular value accorded to time was first, and primarily an economic value, or a value that could easily be translated into economic terms. It was not only the fact that the cognitive representation of the act of production presupposes an ordering of activities in a sequence. This alone would not have been sufficient to enhance the value of time. When the 'multiplier effect' of the machine was added, time came to be seen as a productivity factor and hence something that was – at least in principle – bound to become scarce. Time became the medium in which human activities, especially economic activities, could be stepped up to a previously unimagined rate of growth. The account of social historians in the early days of the industrial revolution are very telling in this respect and vividly demonstrate the rapid emergence of a whole new set of attitudes vis-à-vis time. Time became a symbol for the production of economic wealth and was treated like a valuable object in itself, as Benjamin Franklin so bluntly put it in his advice given to a young tradesman. Time could be acquired by an individual like any other object; it could be saved or unwisely spent, and was to be invested properly, so that it could not get lost. The notion of progress that cropped up in this connection pointed to the possibility of unlimited wealth and betterment of human life in future. Time was no longer considered to be something sacred and given, at best reproducable trough the "myths of eternal return", as Eliade has described them, but became equated with an economic object, whose production it symbolized; it became possible to 'make more out of one's time'. Timekeepers, according to Rezsöhazy (1970), became at that time the regulators and control-

44

lers of action. Work was paid according to the time spent at it, and timekeepers were the quantifiers which transformed an activity into its monetary value.

Today, in most industrialized societies, time is highly valued and considered to be more scarce than ever. Scarcity of time is on the verge of replacing the scarcity of money, not only for those individuals who have enough money already, but for the economic system of whole societies. Economic activity is increasingly measured by the number of hours it takes to produce certain goods, and this serves as an apparently sound basis for comparing the economic standards of different countries. Every sense of crisis, so often connected with the fear of scarcity, also induces in us a feeling that 'time is running out', as though it were limited, just as money used to be before inflation became rampant. If one inquires into the reasons why time has become scarce and highly valued, the answer is this: time has become the medium in which the results of production may accumulate. Yet, a strange, complementary relationship seems to exist: the more one wants to produce, the more time becomes scarce. Those who produce most have the least of it. It is also no longer necessary to limit time as a medium of production to the production of economic goods alone. It has become a medium of production of all sorts of human, especially social, activities. We want to get to know more and more people; we want to do more and more things until recently, we were led to believe that we live in age of abundance economically-speaking as well as with regard to social activities. So the question arises: is it possible to increase the amount of time available in order to produce more? Is it possible to 'produce' time?

Measured against any kind of external reference scale it is quite obvious that the amount of time available to an individual or a society can neither be increased nor decreased. It is possible to produce *in* time, as a medium in which activities take place, but not time itself. And yet, the desire to produce more and to accomodate more and more activities in the time span available has led to a curious phenomenon: we appear somehow to 'borrow' time by extending our time horizon into the future. As there is a surplus of wishes and desires, plans and activities that are to be realized and that cannot all be accommodated in the present, the temporal horizon is widened by extending it into the future. We 'borrow' time by living already partly in the future. It thus becomes clear under what conditions time is experienced as something which is scarce. Faced with a great number of possibilities, activities and plans that are to be realized, the temporal medium of present time appears as insufficient, hence scarce. It is therefore the discrepancy between the relative abundance of possibilities and the relative scarcity of present time which determines this particular experience of scarcity. In traditional, non-industrialized societies these problems do not arise. Time appears as given on the societal level, where it is generated through certain macrotemporal, often cyclical activities. Since such activities cannot be altered by the individual, available time will neither increase nor decrease. However, there is a parallel to the wish to expand the present: time, as the medium in which activities are to be realized, is not extended by 'borrowing' from the future, but is being 'reproduced' by bringing back elements from the past that have been encountered before. It may well be that the myths of eternal return offer an opportunity similar to the one which industrialized societies have created by borrowing from the future, namely, to transcend the immediate present. It has been maintained that to speak of a present makes sense only because there is the individual experiencing it subjectively with his consciousness (Fraisse). To this we might add, that to speak of societal past and future makes sense only because a society or group cannot accomodate all its members' social consciousness in the immediate present. If our assumption is correct, that an extension into past or future occurs because the present cannot accomodate more than a certain share of possibilities, we must then ask why it is that some societies extend their present into the future, and why others extend it into the past?

We have seen that under those conditions in which there exists the prospect of a great number of possibilities to be realized, present time will be experienced as being relatively scarce, and time will be considered as a medium in which at least some of the perceived possibilities can be actualized. In societies where activities and economic goods can be produced more or less at will, where time is bound to become an extremely scarce 'commodity' that can be increased only be opening up towards the future. This is achieved by partly living in the future already, by planning it, and through attributing linearity to it. This extension towards the future has its roots in the need for more time, which is in turn a necessary pre-condition for the production of more goods and activities. By contrast, in a non-industrialized society, the possibilities of producing goods and activities do not exist in abundance, nor is production perceived as itself 'producable'. Hence, time will not be experienced as being scarce, for in relation to what should it be scarce? However, in non-industrialized societies, time is also regarded as a medium, but not one which serves mainly to produce economic goods. It is a medium in which social relations are to be created, maintained and structured. These relations exist among humans as well as between individuals and divine beings. They are structured in a temporal sense insofar as great importance is attached to the performance of acts at the right moment, of knowing not only one's place in the social structure, but also 'one's time' in the temporal structure of the society. It may be as important in such a type of society not to be too fast, not to be too much ahead of each other, or not to interfere with the inter-generational cycle, as Munn has shown for the Trobriand Islanders, as it is for us to work harder, in order to have more money with which we can do more things.[2]

Temporal Strategies: Accomodation and Displacement of Social Needs and Claims

In our inquiry about the mechanisms of extending the social present into past or future, it is useful to introduce a distinction here which is well established in sociology. I will describe this distinction in an over simplified way. If we ask what it is precisely that human beings want to accommodate in the social present, the answer boils down to three issues: (i) the satisfaction of economic needs, which is largely accomplished through the production, distribution and consumption of economic goods; (ii) the satisfaction of needs for social recognition and prestige that enables an individual to acquire or hold certain positions in a social structure; and finally, (iii) social claims with regard to power, i. e., control over one's fellow men and their activities. These are, in a nutshell, the issues usually summed up under the ideas of class, status and power. It is a truism that a social structure usually does not secure equality for its members in these three domains, for only a fraction of the members will be able to satisfy their needs in a given social present. The others can either resign to their fate of losing out, or they will tend to strive for a higher degree of satisfaction. Another possibility would be that they are granted more satisfaction partially, i. e., in certain periods of their lives, like in their economically productive years. It is impossible to discuss here in detail how the temporal accomodation of needs and claims can be accomplished, and what results different structures of inequality in this respect might have. Instead, we have to concentrate on a few consequences of different ways of accommodating some of these needs and claims in the temporal structure of a society.

Let us begin by inquiring about the conditions under which the social claims to power are displaced from the social present. It is a peculiarity of such claims that they need a legitimizing basis, often located in interpretations of the past, or systematically connected with it, through laws that are to guarantee their continuity. Only power which needs no further legitimation because it is based on coercion exerted in the present, is exclusively rooted in the present. It can be observed that the past of any society is usually arranged in such a way that it can serve to legitimize and support the power claims of those groups or individuals that dominate in the present. If these claims are challenged, if a new group or an individual acquires power, or if other major shifts in the political configuration occur, the past is likely to be re-arranged accordingly. If, on the other hand, there is no room in the present social structure to accomodate the power claims of a new group by assigning them the positions they wish to attain, this group is likely to find itself also prevented from establishing a legitimizing basis in the past. One strategy for such a group may be to abandon the placing of their claims in the future, for they have to skip the present. This may be accomplished, for instance, by referring to certain features of the society's past, claiming that they are neglected in the present and demanding or promising that they will be re-installed in the future. This is a strategy commonly found among revitalistic and restorative movements. Another possibility is to adopt a new course which negates and devalues the past and is thus explicitly directed to the future. This is a strategy found among utopian and radical movements of all sorts, as they want to cut themselves loose from the past and attempt to create a New Man in a New World. One of the first historical records for the displacement of power claims by a group that lost out in the present may be found in certain writings of the Near East from 300 B. C. onwards (Holl). In these quasi-official writings, the idea of an after-life in which distributive justice takes place appears for the first time. It can be shown that these ideas originated in a class of scribes who lost political influence at the time and who apparently took revenge by painting a future modelled after their ideas. Similar displacement strategies into the future have been observed in millenarian and revolutionary movements, as rich literature on the subject testifies. The common pattern shows that whenever social claims to power cannot be enacted in the present, a strong tendency exists to transfer them under various forms into the future. Under such circumstances the future also becomes the basis for legitimizing present political action.

An interesting case for displacement into the future is that of the charismatic leader who builds up a political following. Such a leader who may be an empire builder or chief of a tribe may lack the resources to put his plans and visions into practice. He may be observed to embark on a course of action in which he trades the resources possessed by his followers for future promises of rewards (Bailey). This is an intricate game of strategizing in which an immediate reward for support offered by the followers is not possible, so that the charismatic leader has to offer future rewards. It therefore becomes necessary for him to construct a vision of a future, in which the promises that he makes occupy a credible place. Thus, the future becomes a surrogate for a present.

Let us now turn to the satisfaction of economic needs and claims of social recognition. As in the case of the charismatic leader, we shall encounter again an interesting mixture of scarcity of resources – symbolic and material – and their interconnections, as the basis upon which displacement occurs. In any traditional society the means of economic production are, in general, very limited; this means that the satisfaction of economic needs has to be largely accommodated through immediate consumption in the present. At all times, it has been possible, however, for a small group to extract an economic surplus which was invested by those in a position to do so, in ways that would strengthen their own claims to power and/or their claims to social recognition (B. Moore). The relics of the resulting conspicuous consumption can still be admired in many parts of the world today, where palaces, temples and cathedrals were not only built to serve the glory of the Almighty, but also to perpetuate the social prestige of the persons of highest status. This brings us to the question of status preservation. The scarce factor in this case is that of symbolic resources, which have to be kept scarce if they are not to lose their symbolic value. Even in traditional societies where status is said to be largely ascribed and not achieved, strategies must be employed in order to preserve positions of high status, as these are constantly threatened by others who also want to attain high status. This need not necessarily occur in an outright, status-competing fashion. In the rigid Indian caste system, where status is normatively prescribed and strongly enforced through religious sanctions, threat of status decline in the highest caste appears in the form of increase of numbers of their own members (Douglas). The Indian Brahmins are led to practice a form of population control largely for reasons connected with status preservation and the intricate ways of manipulating status symbols. Among other groups having high status, one can also often find a pronounced concern with safe-guarding a highly traditional and status-maintaining way of life. What this means in terms of time-structuring will be discussed below.

Time and Symbolic Resources: Long-term and Short-term Perspectives[3]

It was argued above that those who enjoy positions of high status tend to work to preserve their privileges or power for the future. This can only be accomplished by developing strategies with regard to the preservation of status or power. Such strategies in their turn necessarily depend on and generate long-term time perspectives. On the other hand, we also observe that many consumption needs do not require a long-term perspective, as the need vanishes upon being satisfied (though it might return later). Before we can explain how these strategies intermix, we must briefly discuss how status is gained in the first place and how economic goods and status are interrelated.

We start from the assumption that every society consists of a system of unequal positions, that is, some positions are more highly valued than others, and are endowed with different privileges, rewards, and recognition. Status systems appear to be universal phenomena; generally, the status that an individual holds is only partly dependent on the positions where he finds himself, and partly determined through the objects with which he affiliates himself. Objects are to be interpreted here in a broad sense – they can be persons, goods and even ideas. When an individual associates himself with an object that is highly valued in a particular society or group, symbolic resources are put at his disposal, and he will gain in status. In case he affiliates himself with an object which is negatively valued, he depletes himself in symbolic resources and his status will decline. Nor does the valuation of objects remain static. It is subject to the dynamics of social valuation which itself depends partly on affiliation with other objects or persons, and partly on the frequency with which they appear. Rare objects have great symbolic value, as their information content is high (Schmutzer, v. Foerster). The probability that they will be co-opted by a person of high status is, therefore, also high.

Thus, we may observe that an abundance of symbolic resources and of objects, carriers of symbolic resources, leads to their inflation. They offer little attraction for those who want to preserve their high status position, and will therefore either be passed on rapidly or excluded from affiliation with high status persons. This is the result of a very general process of symbolic devaluation that occurs whenever objects with symbolic value, even if this value is initially high, tend to lose it by becoming more frequent. Only rare objects have high symbolic value that can be preserved at the price of keeping them rare.

Based on these considerations, we can now return to the question of how long-term and short-term perspectives are associated with different positions in the social structure. The upper strata, generally composed of persons occupying high status positions, must employ strategies directed to two goals. First, they must find rare and potentially valuable objects with which they can affiliate themselves (search strategies); secondly, they will tend to employ strategies that permit them to keep those objects rare that have high symbolic, status-conferring value. Such strategies lead to various devices through which the symbolic resources can be manipulated. But the creation of strategies necessitates the development of long-term time perspectives. In a situation of economic surplus production, consequences will arise for symbolic resources also. As more new goods appear on the market, the rarity of objects monopolized by the upper strata will constantly be threatened, as these objects will filter down gradually to the lower strata and become widely distributed. The upper strata thus will develop an interest in finding or producing novel objects with high symbolic value with which they can affiliate themselves first.

But, as the process of economic surplus production is not to be halted, the lower strata will be induced to develop a constant desire to obtain new objects which also will confer status on them. This can be achieved by the development of long-term perspectives with regard to the means with which to obtain such goods. The typical case is described by sociologists as that of rising expectations, i. e., the belief in betterment not only in terms of the material existence, but also with regard to a betterment of their status position. The attitude manifests itself in new aspirations for higher education, and also in other ways in which industrialized societies have normatively prescribed means of status achievement. But the lower strata will always lose out in such a process, as long as the upper strata are in a position to manipulate symbolic resources in their own favour. Such manipulation is achieved by keeping status-conferring objects which, at least for some time after their initial appearance, will tend to remain rare.

We can specify now under what conditions non-induced, originally short-term perspectives are likely to be developed in different strata of a society. This is generally the case in all those situations in which there is either no feasibility or no necessity to develop strategies for attaining or maintaining status-conferring symbolic resources. The latter case of no necessity arises among the upper strata, whenever their power or status positions (which we always assume to include the possibility of extracting economic surplus from the lower strata and of converting it into either power or status) are held to be so secure and unchallenged that no need is felt to be concerned with its preservation. To be sure, even in highly hierarchical societies this will happen only for short periods and never wholly so.

The lower strata are left then with short-term time perspectives, as they do not perceive, nor do they objectively have a chance to develop reasonable strategies in order to obtain convertible symbolic or material resources. Only in those circumstances where the means of status gain are prescribed within their own group (as is the case where status could be gained, e. g., by having a beautiful funeral or a lavish marriage ceremony) can we expect to find long-term time perspectives. With regard to a status gain relative to, and often at the expense of, the upper strata, however, conditions generally are such that strategies are to no avail, hence not likely to be developed. Thus, it appears that the lower strata have only two possibilities open to them, if they want to develop longer time perspectives. Either they can not adopt the long-term perspectives induced in them by the upper strata, thereby voluntarily or involuntarily taking part in the process of symbolic devaluation which works to their own disadvantage; or, they can develop long-term revolutionary strategies, whereby they place their claims to increased status into the future. In content, such revolutionary strategies amount to attempts to abolish the rarity of status-conferring goods, or simply, to occupy the positions held previously by the members of the upper strata. It is, however, difficult to visualize a society in which the rarity of status-conferring goods is completely abolished. If new goods are introduced, they would have to be introduced in great numbers and, preferably, also at the same time. These are conditions which are difficult to meet, as it is doubtful whether productive and distributive mechanisms in any society can be synchronized to a degree that would eliminate time-lags in their distribution. Such truly revolutionary strategies would amount to the abolishment of value-creating mechanisms, at least as they are now controlled by those who have high status. However, while probably no society can exist without some mechanism which produces symbolic resources, the invention of more equal production mechanisms does appear possible.

Abundance of Economic Resources:
Prolonged Present and Shortened Future

After this excursion into the realm of the production and management of symbolic resources and their impact on short-term and long-term perspectives, we want to return to the condition which characterizes most our present-day industrialized societies: the relative abundance of economic resources. It was explained earlier that in a situation of a surplus of resources the future, and the extension of the present into the future, acquire new meaning, for it holds out the promise that more goods may be produced and more activities can be performed than the present can hold. We have seen that this is likely to lead to an extension of the present into the future, into which plans and expectations can be laced. Future possibilities are investigated in more and more detail, bringing the future closer to the present. Compared with the vast domain of possibilities that the future contains, the present appears to shrink to a short interval which passes with increasing rapidity. The future is no longer seen as a mere extension from past and present, as was the case in a situation where resources were still scarce, but turns instead into an open-ended, linear progression which holds out the promise of more, better and even 'denser' times. In short, we are describing progress. A common way of experiencing this conception is expressed in the readiness to 'believe' in the future and the willingness to exchange the present for it (Bell & Mau).

Yet, at almost the same time, one can observe the appearance of countervailing tendencies which result in shortening again the long-term perspective that has just been developed. The future comes to be seen no longer as an extension of the present, but rather as an already overloaded present, multiplied as it were by some growth factor. The reasons for this change in the qualitative nature of the future seem to be the following. The preoccupation with the future as an extension of an overloaded present leads to a closer interconnection between the two, at least with regard to the near future. The extension of the present into the future also leads to an extension of this future backwards into the present. The more the future is planned, the more it is filled up with activities, wishes, plans and desires, the more it comes to resemble the overloaded present which it was intended to expand. In such a situation new modes of perception are likely to be developed which no longer permit a clear-cut distinction between present and future. As with a busy man's schedule that has been already filled up for months in advance, a society intensely preoccupied with its future is likely to find itself 'booked up' in advance and thus has to learn to re-code its present. Thus, we have a two-fold process at work. First, the present becomes extended in its boundaries by 'borrowing' time in which more can be produced. This leads necessarily to long-term societal perspectives. However, as the future is filled up by planning, new boundaries appear that mark off a more extended present from a future that fuses with it. This fusion results in a new kind of extended, yet short-term societal perspective.

In this connection it is instructive to cite a finding of a survey conducted by Galtung (1970) in several countries on the images of the year 2000. It was found that in highly developed societies the year 2000 is seen as something far away, while the prognosis of what the general state of affairs will be then was a highly sceptical one. By contrast, in developing countries the year 2000 was seen as being relatively near; it was believed that a rapid rise in economic development will be reached by then. This may not only had be interpreted as a typical case of rising expectations, but fits also well into the pattern described above, illustrating the two-fold process at work. The developing countries find themselves at a stage of the process today where a long-term perspective has just been acquired, made possible by a newly experienced situation of relative abundance. The developed countries find themselves at a stage where they have already acquired a conception of the future as a newly extended present, one which necessitates new cut-off points experienced as (new) short-term perspectives. It is very probable that the re-coding of the present is facilitated by the fact that the extended present appears much more problematic than was formerly believed. It would thus require a higher degree of attention and more intensive information processing on the societal level. These requirements are also likely to lead to a shortening of the time span which can be attended to. Some of the problems, formerly believed not to occur before the year 2000, are with us.

On Timekeepers and Time-measurement

The process of symbolic valuation described before can be taken as paradigmatic for the changing symbolic value accorded to timekeepers over the ages. As the possession and use of clocks filtered down from the aristocracy to the bourgeoisie, to be later almost forced upon the working class, their history came to illustrate the processes in which persons or groups affiliated themselves with timekeepers as status symbols. Likewise, the introduction of clocks to non-European civilizations, such as those of China and Japan, where clocks remained for centuries the playthings of the wealthy and the mighty (Bedini), illustrates the same process. In the modern world where economic production necessitates time-keeping and facilitates the wide-spread possession of timekeepers, clocks have been replaced by watches. The formerly communal symbol, prized for its rarity and affiliation with high status persons, including those inhabiting the heavens, has yielded its place to utilitarian, functional objects that are individually owned and are in practically everyone's possession. According to our hypothesis, such a mass-diffusion process is invariably accompanied by symbolic devaluation, at least of the object that has been chosen to represent the concept of timekeeper. But if timekeepers can no longer be represented by mass-produced watches and be a highly valued object at the same time, where is the new status symbol of a timekeeper to be found? Apart from fancy variations of the product, it might be that the close scheduling itself, and perhaps the human operators who serve it (such as secretaries or service institutions) will replace the clock as status symbol. The modern high status person is one whose time is extremely valuable and who, correspondingly, possesses very little of it. He has to spend it wisely, distribute it economically among his many activities. There is also a concept of time-economizing behind this pattern; time which can be saved can also be spent elsewhere, presumably at more rewarding activities or with more rewarding persons – those that may serve to increase one's social status even more.

To have too much time at one's disposal has, paradoxically, become an indicator in our society of having nothing of importance to do and, consequently, of being of no importance. The concept of time behind such an attitude is no longer based on the movement of stars, planets or gods, and their meaning for human beings, but reveals rather a time concept based on economic activities and how they can be transformed into various social amenities and privileges. We have moved towards a state where man keeps, as it were, his own time. But, specifically, whose time? Should the high status person's time or the busy man's time be more valuable than that of anybody else? Is there no common time left in which everybody can share and participate – a communally experienced and valued time?

At this point we want to return to the questions raised at the very outset of this paper. To what extent do the measurement scales that we use to measure time and the underlying conceptions of dimensions on which they are based, influence our notion about time? If one compares today's time-keeping operations, especially in measurement aspects, with those of former times, one difference emerges very clearly. In former times, different scales were used for keeping or measuring the times of different activities. In a non-scientific world all scales, including those that were used to measure time, are adjustable to circumstances. It appears to be a peculiarity of a scientifically-minded society, however, to prefer scales which are unambiguous, exact and universally valid. As Leach has pointed out, we seem to operate today with scales of great precision but of fewer dimensions than in earlier epochs. The reasons for this are to be sought in our preference for all values to be expressed in interchangeable numbers. Just as in economies the value of different products, of food, labour, or land, can be expressed in terms of a single numerical dimension (quantities of money), so we can now convert formerly different time scales and the activities that were measured by them into one standardized, uniform and uni-dimensional time scale. This trend may be interpreted as the result of an increase in complexity on a large and broad scale of social organization. Yet, the question remains – and it is an open question which I want to leave open – whether this does not imply a loss of richness in structure, and a loss of the creative power that lies in differentiation. To set up multiple scales destined for different occasions and purposes poses no difficulty whatsoever from a mathematical or technical point of view – multi-functional and multi-dimensional timekeepers could easily be constructed. If this would happen, human activities would have to adjust again to different circumstances and take on a newer and richer quality, as a consequence. It would be another example for the interrelation between time structuring and time measurement that we have observed throughout this paper. Only this time, time measurement would re-influence the structuring of time. As Bourdieu (1972) put it when he referred to the cycle of social relations generated through gift-giving in North Africa: to abolish the interval is to abolish strategy. I would like to paraphrase him: to abolish a uni-dimensional time concept, is to restore the richness of social life.

There is a certain addictive quality in the preoccupation with time. Like any true addict, I encountered others whose visions I was fortunate to share and whose intellectual generosity I wish to acknowledge here: Edmund Leach, Otthein Rammstedt, and especially Manfred Schmutzer.

References

1 The literature on the subject is vast. See especially Moenks (1967) and Luescher (1970).
2 This point may be illustrated by the following quotation: "I'm working so hard that I'm killing myself and wrecking my family, but I'm making so much money that I can afford it." T. C. Schelling, Foreword, Symposium: Time in Economic Life. Quarterly Journal of Economics, 4 (1973), 627.
3 The following section is heavily indebted to the work of Manfred Schmutzer.

Bibliography

Barndt, R. and Johnson, D.: "Time Orientations in Delinquents." Journal of Abnormal and Social Psychology, 51 (1955), 343–45.
Bailey, F. G.: Strategems and Spoils. Oxford: B. Blackwell 1970.
Bedini, S.: this volume.
Bell, W. and Mau, J., eds.: The Sociology of the Future. New York: Russell Sage Foundation 1971.
Bernstein, B.: Class, Codes and Control. Vol. I. London: University of London Press 1971.
Bourdieu, P.: "La société traditionelle, attitude à l'égard du temps et conduite économique," Sociologie du Travail, I (1963), 24–44.

Bourdieu, P.: Esquisse d'une théorie de la pratique. Genève: Librairie Droz 1972.

Deffenbacher, K. and Brown, E.: "Memory and Cognition: an Information Processing Model of Man." Theory and Decision, 4 (1973), 141 –78.

Douglas, M.: "Population Control in Human Groups." British Journal of Sociology (1966), 263.

Eliade, M.: Myths, Dreams and Mysteries. London: Harvill Press 1960.

Findlay, J. N.: "Time: a Treatment of some Puzzles. In Problems of Space and Time. Ed., J. J. C. Smart. New York: Macmillan 1964. (1941).

Fraisse, P.: The Psychology of Time. New York: Harper & Row 1963.

Galtung, J.: "Images of the World in the Year 2000." Vienna 1970. Private communication.

Galtung, J.: Members of the Two Worlds. Oslo: Universitets-forlaget 1971.

Goody, J. and Watt, I.: "The Consequences of Literacy." Comparative Studies in Society and History, 5 (1962–63), 304–26; 332–45.

Holl, A.: Personal communication.

Holmberg, A.: "Age in the Andes." In Aging and Leisure. Ed., R. Kleemeier. New York: Oxford University Press 1961.

Lazarsfeld, P.; Jahoda, M. and Zeisel, H.: Die Arbeitslosen Von Marienthal. Leipzig: S. Hirzel 1933.

Leach, E.: "Some Anthropological Observations on Number, Time and Common Sense." Paper presented at the Second International Congress on Mathematical Education. Exeter 1972. Mimeo.

LeShan, L. L.: " Time Orientation and Social Class." Journal of Abnormal and Social Psychology, 47 (1952), 589–92.

Lewis, D.: Five Families: Mexican Case Studies in the Culture of Poverty. New York: Basic Books 1959.

Luescher, K.: "The Social Psychology of Time." Paper presented at the Colloquium in Social Psychology at Cornell University. Cornell 1970. Mimeo.

Michon, J.: "Processing of Temporal Information and the Cognitive Theory of Time Experience." In The Study of Time. Vol. I. Eds, J. T. Fraser et al. New York: Springer-Verlag, 1972.

Miller, S. M.; Riessman, F.: Seagull, A. A.: "Poverty and Self- indulgence: a Critique of the Non-Deferred Gratification Pattern." In Poverty in America. Eds., L. A. Forman, J. L. Kornbluth and A. Haber. Ann Arbor 1965.

Moenks, F.: "Zeitperspektive als psychologische Variable." Archiv für die gesamte Psychologie 119 (1967), 131–61.

Moore, B.: Social Origins of Dictatorship and Democracy. Boston: Beacon Press 1966.

Moore, W.: Man, Time and Society, New York: Wiley 1963.

Munn, N. D.: "Symbolic Time in the Trobriands of Malinowski's Era: an Essay on the Anthropolgy of Time." In N. D. Munn: Essays in Social Symbolism (To be published).

Needham, J.; Ling, W. and de Sola Price, D. J.: Heavenly Clockwork: the Great Astronomical Clocks of Medieval China, Cambridge: Cambridge University Press 1961.

Nuttin, J.: "The Future Time Perspective in Human Motivation and Learning." in Proceeding of the 17th International Congress of Psychology. Amsterdam: North-Holland Publishing Company 1963. Pp. 60–82.

Ornstein, R.: On the Experience of Time. Middlesex, England: Penguin Books 1969.

Ortiz, S.: "Reflections on the Concept of Peasant Culture and Peasant Cognitive System." In Peasants and Peasant Society. Ed., T. Shanin. Middlesex, England: Penguin Books 1971. Pp. 322–36.

Pocock, D.: "The Anthropology of Time-reckoning." Contributions to Indian Sociology, 7 (1964), 18–29.

Price, D. de Solla: "Clockwork before the Clock and Timekeepers before Time-keeping." This volume.

Rammstedt, O.: Revolution und das Bewußtsein von Zukunft. Universität Bielefeld 1972. Manuscript, private communication.

Rezsöházy, R.: Histoire du mouvement mutualiste chrétien en Belgique. Paris/Bruxelles: Erasme 1957.

Rezsöházy, R.: Temps social et développement. Bruxelles: La renaissance du livre 1970.

Roth, H.: Time Tables: Structuring the Passage of Time in Hospital Treatment and Other Careers. Indianapolis: Bobbs-Merrill 1963.

Schmutzer, M.: "Social Crystallization: Variations on a Structural Theme." Essex University. Internal Memo, 1974.

Schneider, L. and Lysgaard, S.: "The Deferred Gratification Pattern." American Sociological Review, 18 (1953), 142–149.

Singer, J. L.; Wilensky, H. and Mc.Graven, V. G.: "Delaying Capacity, Fantasy and Planning Ability." Journal of Consulting Psychology. 20 (1956), 375–83.

Sorokin, P. and Merton, R. K. "Social Time: a Methodological and Functional Analysis." American Journal of Sociology, 42 (1937), 615–29.

Szalai, A., ed.: The Uses of Time. The Hague: Mouton 1972.

Talmon, Y.: "Millenarian Movements." Archives Européens de Sociologie, 7 (1966), 159–200.

Toda, M.: "Time and Space in the Structure of Human Cognition." This volume.

Vercauteren, P.: Cahiers d'observation de paysans ivoriens. Louvain: dactylographié 1965.

Von Foerster, H.: "On Self-Organizing Systems and their Environment." In Self-organizing Systems. Eds., M. Yavits and S. Cameron. Oxford, England: Pergamon Press 1960.

Worsley, P.: The Trumpet Shall Sound. London 1968.

Knowledge and Its Application:
The Context of Science and Technology

The Information Society: Its Impact on the Home, Local Community and Marginal Groups

Information Technologies: Revolution or Continuities?

When attempting to assess the effects of a new technology on the social fabric, our commentaries as prophetic contemporaries of an unfolding development are not only marred by the well-known difficulties and inadequacies of forecasting models. Rather, and more serious in my mind, are the conceptual shortcomings which have plagued the understandig of the relations between technological change, economic production and value shifts in society. Used as we are to fall back on extremely simplistic cause and effect models – in which very often technology is projected as an autonomous force causing disruptive or benevolent changes on its way to impinging upon society – our conceptual apparatus, our collective mode of understanding of the subtle processes of interactions remain underdeveloped. The very fact that every technology encapsulates and embodies social relations, that 'artefacts have politics' in the pungent phrase of L. Winner, is all too easily overlooked, hidden behind the hard-ware appearance and the glittering facades of shiny new designs and gadgetry in which modern technology presents itself to the customer. Work relations, the daily routine interaction with the machine and with a technological system, and mind relations, the imperceptible forms of adapting our mode of seeing und thinking to the exigencies of that system, usually remain nameless and hence collectively anonymous: the 'problematique' of the interface between society and technology is acknowledged only when it transforms into 'problems', be it in the more dramatic form of resistance and protest against a new technology, be it in the more furtive, yet no less alarming prospects of future unemployment, health and safety hazards or the cumulative effects of a potentially de-personalizing environment on the individual. The neglect of the problematique and the belated acknowledgement of newly emerging and pressing problems leads to the familiar oscillation in assessment. Technology is simultaneously credited with the potential of its positive applications and decried for its equally plausible negative and

Reprinted with permission from Bjorn–Andersen N., Earl M., Holst O. and E. Humford, Information Society – For Richer, for Poorer, Commission of the European Communities, Fast Programme, North Holland Publishing Company 1982.

destructive effects. This ambivalence, fixed upon an end product and its use, serves to hide the social, political and economic forces which shape any innovation process and impinge upon the process of embedding a new technology into the social context. Not only is it extremely difficult to isolate the introduction of microelectronics (or any other technology) in either product or process form from organizational or other changes accompanying it, but the emphasis placed upon counterbalancing positive and negative effects obscures the social processes of disruption and adaptation necessary to accommodate a new technology. Altering the social conditions and transforming them into the preconditions for the successful implantation of a new technology is no simple cause-effect operation: rather, it is distorted and a painful historical process, interpunctuated by shifting patterns of social conflict.

Many of the commentators on the new information technology have stressed its 'revolutionary' qualities. The stupendous success of the silicon integrated circuits, both in increase of complexity, lowering of costs and concomitant boost of sales figures; its universality and ranges of applications, the convergence of several related peripheral industrial innovations – all well documented in figures and forecasts – have been hailed as the latest feast of 'Unbound Prometheus'. The heralded information revolution at the work place and especially in the office, in the home and local community, on a word-wide basis with truly staggering and most likely sad consequences for the developing countries, may well be on its way. Yet, from a sociological perspective some strong lines of continuity appear as well. Rather than joining the ranks of those who project forecasts into the future, I wish to examine the arrival of microelectronics and of the Information Society as the accomplished projection of historical trends that have been endemic in the development of Western industrialism under capitalism.

Rationalization Continued

The first string of continuity which I see is embodied in the tremendous potential for rationalization and productivity growth that microelectronics offers. Rationalization, both in its technical component (i. e. the substituting effect of the machine in lowering the material and labour input) and its economic component (lowering capital input, lowering costs and increasing the profit rate) is an integral part of technological progress. It consists of changes in production methods which allow more output to be produced from a given volum of labour and resources, or allows a given output to be produced with a smaller volume of labour and resources. The productivity increasing impact of technological change in the Western economies since approximately 1880 has been intimately linked with the rationalization of production methods (Rosenberg, 1974). What has become known as Taylorization and Fordism, as scientific management, human and social engineering, so aptly described by David Noble as the concurrent emergence of the rise of corporate capitalism in the US and modern technology with its inherently rationalizing elements (Noble, 1979), is but the mechanical predecessor of today's electronic rationalization process. It is no coincidence that the social site of the former was the shop-floor, while it is now becoming the office. If it is correct that the typical office worker in advanced countries of the West is supported by only $ 2.000 worth of equipment in comparison with 15 – 20 times that amount for factory workers (Rada, 1980), these figures signal the tremendous potential for productivity growth in the information sector of the economy, while the further growth in the industrial sector has become one of diminishing returns.

Another striking parallel illustrates the commonality of one of the principles of rationalization embodied in technology. Ford's ingenious discovery of the conveyor belt consisted in reversing the 'order of things' by keeping people motionless and moving the pieces to be produced. Likewise, information technology makes it superfluous for people to move in order to traverse distances by bringing information, which is produced, processed and stored at a distance, to them. For all we know and what economists tell us, there is no reason to doubt the rationalization potential offered by microelectronics, which in the words of the author of an OECD report will make 'the electronics complex . . . the main pole around which the productive structures of advanced industrial societies will be reorganized' (Norman, 1981). Nor does it come as a surprise that the main apprehensions have centered upon the labour displacement effects of the new technology and related to it, changes in the qualification structure of the labour force. While in the past, the de-skilling of workers (Braverman, 1974) and even the slow erosion of the working class itself (Gorz, 1980) has been accompanied by the rise of a managerial class of office workers, albeit stratified in itself, it remains an open question to which I wish to return later, how the de-skilled part of the population will be accommodated and what changes in the stratification system of our societies are to be expected.

Expanding the Techno-system

The second string of continuity lies in the systemic nature of the new technology, both in its conditions of emergence as a science-based technology and in its diffusion and application. Ever since the systematic harnessing of science for the purposeful and economically profitable purposes of continued innovation which began in the last quarter of the 19th century, the spill-over effect of the convergence of several, equally science-based fields of technology application has multiplied, allowing the kind of innovation clusters that Kondratieff and Schumpeter have analyzed. Likewise, the systemic nature of a mass-extended economic and technical infrastructure has become a feature of the way in which modern technology is diffused and used. The building of the necessary infrastructure, from the railway to the electricity grid, is but one illustration of the emergence of techno-systems which have engendered a host of legal and pricing regulations, standardization norms and concomitant institutions in their wake. The main effect of these techno-systems on our theme is their inclusive nature: once the techno-system is about to establish itself, complementing or superseding the older, more fragmented and incomplete technical application fields, it becomes virtually impossible to remain outside its gripping force. This holds for the iron laws of competition on a world market, but equally for the consumer. The power of the techno-system consists in creating numerous internal interlinkages and new chains of dependencies. From all we know, microelectronics promises to become the most inclusive of the techno-systems we have witnessed so far. To remain outside not only carries heavy penalties, but amounts to exclusions from a society whose work and mind relations have become even more intimately linked with bureaucratic and institutional regulations than was ever the case before.

Making the Human Operating Unit Reliable

Another powerful continuing line of development within modern technology is its inherent striving to withdraw as many subjective elements as possible from production (manufacturing or office) and – in addition within the scope of microelectronics, – from human information processing in general. This withdrawal has several components ranging from the theme of domination of nature, including human nature, which is inherent in science and technology, to economic considerations of the relative slowness of humans and the expense of labour costs to outright political struggles of taming the labour force (Rada, 1980) (Noble, 1979). Let me quote the frank words in which F. de Benedetti, Managing Director of Olivetti, has expressed this tendency:

"The taylorisation of the first factories, developod as the answer to competiton between companies, is a 'digitalisation' of the productive process. At first, it enabled the labour force to be controlled and was the necessary pre-requisite to the subsequent mechanisation and automation of the productive process. In this way taylorised industries were able to win competition over the putting-out system.

64

Data Processing is therefore a continuation of a story which began with the industrial revolution, which incorporates the development of abstract terminologies within the development of technologies. Information technology is basically a technology of coordination and control of the labour force, the white collar workers, which taylorian organisation does not cover." (Benedetti, 1979)
While I need not further dwell on the theme of 'people are trouble, but machines obey' (The Engineer, 1978), which has accompanied in an often painful way the introduction of technology into the lives of working people (Salomon, 1981), microelectronics makes it possible to extend the elimination of the subjective factor also into realms remote from the familiar struggles between capital and labour. In view of the staggering information processing and storage capacity that the new technology offers, the human operating unit, as it is called, is irreversibly and firmly placed at a distance from what becomes an independently operating data producing, processing and using system. As with labour struggles before, the main question will become one of control and will be a severe test of the strength of our democratic institutions.
The tendency to eliminate the human factor and to create techno-systems that function independently of human unreliability as a continuous theme of technological development carries the self-disciplinary character that has been a hallmark of the civilization process (Elias, 1976) one decisive step further. The techno-system, if it is supposed to interact successfully with the human operating unit (H. O. U.) and to carry out its information processing tasks efficiently, has to be based on certain assumptions as to how the H. O. U. functions and, on a more philosophical level, what the human being is like. Eventually, this reified image of the human being acquires the compelling force of a new social norm, to which actual social beings will have to learn to adapt in an approximation process[1].

Profit and Losses

The fourth and last string of continuity with which I will deal here is the uneven and unequal nature of any technology, once it leaves the sheltered environment of the laboratory and meets with a society which in itself is beset with inequalities. The sheer amount of capital required for the successful R & D and marketing operations of an increasingly capital intensive technology is destined to bring profit to those who have the necessary investment capital and have played their game properly (Norman, 1981). But we also have to consider the process of diffusion as a process in time which cannot unroll simultaneously. The economic history of Western Europe can be rewritten as one of regional imbalances. The reverse of the process of modernization is that of regional pockets of backwardness, created by the unequal course modernization took, both on the continent as a whole and within nation-states. Today, we witness again de-industrialization taking place in some parts of Europe, while the new industries are establishing themselves in more favourable locations. This is likely to reinforce already existing regional imbalances with all the problems that an uneven supply of basic social and health services, and demographic anomalies, brings with it. While not new in itself, the massive introduction of the new technology is likely to create equally upsetting effects in regional imbalances, if no counteracting measures are taken in time.

Another effect likely to continue is the creation or reinforcement of new or already existing inequalities in society via the new competences and skills which are demanded by the new technology. This is not just a matter of equal access to education, but a dynamic process in which advantages will accrue to some, who possess already a considerable amount of educational capital and who will be able to capitalize on the new opportunities, and others, who will be disproportionately disadvantaged by not possessing or by hardly being able to acquire the newly demanded range of competence and skills. Since marginal groups are one of my main themes, I shall return to examine the conditions which make for their emergence.

By way of concluding this introductory section, let me emphasize again two points: despite the seemingly 'revolutionary' features of microelectronics, I believe that this metaphor is misleading in so far as it cuts itself off from the historical continuities of modern Western technology and its embedding into our societies. These major continuities – the ongoing and accelerated tendency towards rationalization inherent in technology; the tendency to disassociate the production process and, more generally speaking, the ways in which we change the human environment, from the human factor; the systemic, all-encompassing, nature of the socio-technical system which has become the condition sine qua non of modern technology and which is compelling in its inclusiveness; and finally the unevenness inherent in the process of introducing a new technology which serves to reinforce already existing inequalities or to create new ones – let the microelectronics revolution (while perhaps justified in a narrow technical sense) appear as yet another culmination in the process of disruption and adaptation which has been a constant by-product, a concurrent syndrom of the ways in which our societies have evolved and undergone change. The main difference I see between former technological leaps and the shift towards an information society is the increasing speed with which this development takes place and the stepped up intensity, if not to say brutality, with which its main social features, as outlined above, are implanted into society. The options open to us do not refer, I am afraid, to whether we want to have this new technology or not, nor is there time or a realistic chance to shape the new technology in the direction of making it more 'human'. The options, as I see it, are largely reactive, in the sense of how to cushion the likely disruptive effects and to invent new strategies and how to integrate a society whose old bonds, the social glue that holds it together, are likely to be put under severe test.

Impact on the Home: The Last Reserve Space

The first site of unplanned disruption which I have been asked to consider is the home: a social space of special significance which has come to signify for us the last sanctuary in a bewildering outside world. But even if we do not adopt nostalgic glasses, it remains a social space in which privileged human interaction in the form of family relations goes on. Paradoxically and in contradiction to the recurrent mourning cries of the 'decline of the family' or even its often lamented death, the family has proven of astonishing resilience, while changing form, content and structure. In today's world, the family is a multi-faceted unit – ranging from the increasingly frequent mini-family of one parent and one child, to the more familiar standard family of parents with their 1,3 children, to more diffuse forms. Due to the increase in longevity, the four-generation family, while not living together, is nevertheless a social reality. Due to more frequent remarriages, a proliferating complex network of relationships binds new members together, while unconventional forms of living together, from the unmarried couple to diverse communal living arrangements, are here to stay.

For all these relationships the home symbolizes the social space which differs in quality from the work place and the instrumentalized relationships that prevail in general 'outside'. It is a space in which repair from the vicissitudes of occupational life still takes place in diffuse and non-professionalized ways and in which that modicum of 'inwardness', of emotional and spiritual recreation is permitted which we call privacy. I say still, because – contrary to the resilience of the family as a social unit of various and diffuse relationships – the home has been successively and successfully invaded by that opposite, the public space. The home has lost gradually in its autonomy, it has been incorporated into a web of administration that regulates how its members are to be brought up, how they are to be cared for when ill or old, how much time, and in doing what, its members are spending outside and how the individual home is connected to the outside world, both in the material aspects of being connected to communal facilities and through the lines of communication with the outside world that have been established via the media. This is not to say that no functions have remained within the home. They are the ones that have mainly been left to women who continue to perform unpaid work within the home in addition to the generally lower paid work they perform outside – caring for the children, the sick, the old and performing that most peculiar form of labour which an increasing division of labour has entrusted upon women, emotional labour. But all these and other remaining functions are essentially a substitute, a reserve, competing in standards and cost with the professionalized services that have established themselves outside the home. Today's home has therefore acquired a touch of a last reserve, a sanctuary perhaps due to the privileged nature and emotional diffuseness of the relationships it may harbour.

But apart from this somewhat romantic vision, the home is the main locus where reproduction takes place. The first major impact of the coming information society which I see, is therefore given through the innate relationship between the sphere of production and its repercussions on the sphere of reproduction. Will the home be able to cope with the increasing stress that many of its members will probably be exposed to when future working conditions are intensified and when the drudgery and harshness of manual labour gives way to the more subtle monotony and psychic stress that accompanies non-manual labour? What about the life-long learning process which has been predicted as constituting the educational pattern of the future, with members being obliged to pursue voluntarily their training in their leisure time. What effects will it have on the quality of home life, if parents and children will have to continue their education indefinitely, the former to keep their jobs, the latter to get a chance to obtain one? The paucity of already available evidence does not permit me to draw conclusions, but it is plausible that the home as reproduction unit which includes of course the restoration of its members' capacity to be active in the sphere of production, will be put under greater stress (Kalbhen, Krückeberg, Resse, 1980). Since many of the presently available mechanisms for coping depend on the presence of women in the home (be it only part-time or in time-off their working hours elsewhere) and their continuing performance of unpaid labour, will they be willing to continue to do so and if not, will they be pushed back from outside employment when the labour market shrinks?

While these are questions of great concern not only to women, which I am unable to answer, my arguments are more definitive on two other accounts.

Firstly, the arrival of the information society will bring a further blurring of the boundaries between work and leisure or translated into spatial terms, between the outside world of work and the home, largely though not exclusively, reserved for leisure activities. The main reasons for this blurring are economic ones and related to the emergence of a self-service economy (Gershuny, 1976) (Skolka, 1976). Due to rising costs for the performance of human services and the comparative lowering of costs at which goods are available, the system of production is said to undergo a decisive change in the sense of producing more goods which can be utilized as a substitute for the procurement of services. It is obvious that this tendency will be boosted by microelectronics. While the economic argument is relatively straightforward, the social implications have largely gone unnoticed. The convergence of work and leisure which is implicitly contained in a fully developed self-service economy is partly due to the industrially produced character of the goods with which the consumer now performs services him- or herself. These self-servicing goods make traditional skills and competence obsolete and demand new ones. The new skills, however, necessary to perform the unpaid work in the self-serving household, are – and this is my main point – commensurate with the mode of work organization which pervades the work life. The blurring of work and leisure in the coming information society therefore rests on two mutually reinforcing developments: the shift to the self-service economy, transforming a number of activities which we now locate in that grey zone between work and leisure, into work, for example, driving, shopping, gathering consumer information on where and what to buy at relative favourable cost, self-service in the household and generally the expanding scope of unpaid consumer work, while work itself becomes more like these activities AND the gradual fusion of the underlying organizing principles. B. Joerges, in a perceptive article, has detailed some of the interesting features of such convergence (Joerges, 1981). For my part, I wish to emphasize the social significance of the increasing resemblance of the organizing principles of work and leisure. The past attempts at subjecting the home under the guiding principles of scientific management have only been modestly successful (Maimann, Loidl, 1980); mainly because factory work allowed the simultaneous division of rather homogeneous tasks, while household chores remained much more heterogeneous and needed to be performed in sequence rather than simultaneously. The massive introduction of self-serving goods into home and leisure activities and the pervasiveness of information technology throughout our daily lives promise a more successful rapprochement this time. Already now one of the unexpected results of this convergence manifests itself. It finds expression in the unspecified longing for an 'alternative' life style, be it in the more conventional and modest form of an exotic place for spending one's vacation, be it by practising Indian meditation techniques or growing bio-food to becoming part of the alternative scene – activities which rest on organization principles distinct from the ones that have come to dominate our work and home life.

The other major impact on the home which I foresee is the growing dependency on information which is produced at a distance. This is not a new theme in communication research at all and it has a recognized commercial and a political side (Jansehen, 1980) (Jansehen) (Reese et al., 1979). It acquires new relevance, however, when one considers the pervasive nature of a telematic communication system with its over-determined supply of information. Many observers have emphasized the potential of the new information technology in overcoming isolation and its community-building component (de Sola Poo, 1980). Others or occasionally the same authors, have equally emphasized the potential for 'personalized' information offered by a dramatically expanded communication system which is based on the venerated assumption of the consumer's rational choice finally before our doors or already in our living rooms (Cowen, 1981). But this amounts to an unresolved contradiction on which the commercial sector thrives: while the new technology reinforces trends towards isolation already existing now, it promises to overcome that same isolation which it helps to generate. A mass-produced and commercialized entertainment supply of information and services advertised to suit the idiosyncratic whims of each indvidual consumer, helps to standardize taste and shape demand according to the guiding principles contained in its supply. But while it is still possible today to escape the market of commercially produced goods and retreat into the do-it-yourself (DIY) corner at least temporarily, the increasing incorporation of DIY into the self-service economy leads to the closing of this escape hatch. Private knowledge and private information – both of which are eminently social since they are based on personal experience and ongoing social exchange and interaction with others – will accordingly be de-valued. Although the celebrated two-way flow of information which the new communication systems promise, opens up novel perspectives, it has to be recognized that communication with one's fellow human beings from private space to private space becomes mediated by a public medium which is either state or industry controlled.

On the political side, apart from the aforementioned control aspects and the extension of the public into the private sphere, an extreme personalization of information processing and communication carries the danger of bringing a loss in competence for collective action which is a vital ingredient for any Western democracy. This leads already to the next site, where we are likely to experience the impact of the new technology: the local community, that intermediate space for social, cultural and political activities.

Impact on the Local Community: The Intermediate Space

Like the home, the concept of community also carries with it the notion of a social space, both in its original sense of a network of relations connecting neighbours and friends outside the home, but in close geographical proximity, and in its later, administrative sense which made the community the most basic level of state government and administration. As an administrative-political unit, the local community embodies a shrinking heritage of de-centralization and of autonomy; as a social concept community occupies the intermediate space, serving as mediating ground between the individual and the family and the more anonymous wider society. The community as an administrative-political unit has retained a number of important functions, like the provision of elementary or even central services in education, health and welfare, to which new or expanded ones are added. A number of attempts are currently under way in Western Europe to de-centralize further and to install community-based services within close range of the potential users and clients. This is done with the cost-conscious intention of utilizing human resources that are typically to be found in the social concept of community: the basic, albeit intermittent mutual support of neighbours, the work of voluntary organizations usually organized on the community level and the vague sense of togetherness which makes people join in efforts to embellish or enrich community life – an ensemble of cultural, social and political activities, essentially voluntary in nature and encapsulating a modicum of spontaneous collective action towards the common goal of enhancing the quality of communal life.

There is an interesting parallel between the tendencies that we have noticed towards the self-service economy and current tendencies in the realm of the provision of health and welfare services, to shift the burden in the direction of the consumer. The self-caring capacities of ordinary people who are the potential clients or users of services provided by the state and the professions are to be incorporated, just as the consumer has to take on additional tasks. As in the case of the self-service economy, a peculiar coalescence of divergent interests can be observed. The economic forces behind the shift towards transforming the consumer into his or her own producer of services meet with the increasing demand – having very different roots – for more meaningful work, for doing one's 'own work' (in German aptly termed Eigenarbeit) and a thriving secondary, black, or shadow economy, based on a mixture to retain or regain some autonomy over one's work and leisure by doing things oneself. It is as though the DIY- movement fragmented, individualistic with an anarchic streak suddenly encounters an industrialized ready-made DIY-economy, ready to devour all individual attempts and illusions about what can indeed be performed for and by oneself. Similarly, under the very real pressure of galloping costs for health and welfare services everywhere in Europe, ways and means are being sought by policy-makers to reduce at least some of the costs. The shift to more volunteer work, the shift towards the community level seems a promising, although in the end probably illusionary answer if costs alone are calculated. In the meantime, at the other end of the spectrum, various self-help and self-care groups have emerged on the community level. While their main grudge turns against what they define as the bureaucratized and professionalized services which are offered to them and which they feel are not covering their real needs, nor their need for more self-direction and responsibility, the two independently arising developments coalesce. However, their encounter is likely to be mutually disappointing, if it does not carry the seeds of serious confrontation.

The tracing of these developments and their peculiar constellation provides important clues for what the impact of information society on the community is likely to mean. The technological potential has been sufficiently spelled out to allow us to recognize the contours. We may expect the displacement of a number of professional services, now performed by humans, by electronically processed computer programmes. Counselling, to name one example, in its psychotherapeutic and more profane facets, can easily be provided by computer programmes with in-built referral systems, if the problem turns out to be not yet sufficiently standardized to be handled by the programme. Reese and his associates in the FRG have investigated the possibility of installing information technology within the provision of welfare services (Lange, Kubicek, Reese, Reese, 1980). Others have explored the community-building potential of information technologies and the possibilities of offering a community-centered information programme. The use of computer programmes for the purposes of diagnosis, of data storage and improved referral systems have been charted out for medicine. Still other potential applications will be forthcoming. Yet, the social meaning of it all is very clear; the rate of diffusion is only subject to the cost factor which in turn will be influenced by the rising costs of human, i. e. professional services: by installing standardized, but in themselves very flexible computer programmes, adapted to the administratively perceived needs of the individual client or groups of clients, self-care becomes an administratively acceptable and in terms of costs, economically viable reality. The only difference will be that this future form of self-care, like the future form of self-service and Eigenarbeit, will have little to do with what alternative movements of today associate with it.

Perhaps this needs a bit more of an explanation. The present discovery of the layman and woman, both on the individual and community level, is fundamentally different, depending on who is discovering whom. From the administrative-political perspective of the policy-maker and politician, the rising costs are staggering and the search for alternatives turns out to be the most attractive, the more they promise to utilize essentially unpaid resources in the form of all kinds of voluntary work. For the clients and users of health and welfare services, to a small extent even of educational provisions, the discovery of the lay person means the partial refusal of professional state-provided services held to be insufficient on several grounds and signifies the reactivation of some degree of autonomy and self-helping capacities. In some cities in Europe, the alternative network has already reached sizeable proportions, although it is largely serving a specific clientele of young people and others who have dropped out or have been dropped out by society. Yet, in its logical consequences there is bound to be open confrontation. The growth and proliferation of alternative self-care movements, while tolerated in the beginning, is bound to violate, once it has reached a certain size and a concomitant growth in self-assurance and internal dynamics, existing norms of professional and legal regulations. It is bound to come into conflict with the monopolies that the professions have erected, criss-crossing our lives, and with the legal provisions, also the result of a long development, which regulate, in the interest of other groups, most of our lives. Just as children have to be sent to state approved schools, sick people have to be treated by state approved doctors. And while the welfare state tolerates the non-use of rights and benefits that it confers, public opinion does not tolerate the most minute forms of abuse (European Centre for Social Welfare Training and Research, 1980). While self-medication and health prevention through the individual is welcome within certain limits, the all-out refusal of the present high-technology based medical system would meet with restrictions and, eventually the forbidding of the more extreme forms of alternative self-cure. In each instance we can, aided by historical examples, see how narrow the margin of tolerance actually is. As long as self-service (self-work) and self-care movements remain relatively modest and do not overstep the niches that a cost-conscious administration of the official services provides, they will in fact prepare the way for the arrival of the new official self-service system based on information technology. If they overstep the boundaries, as they are likely to do, given the internal dynamic of collective movements and the growing divergence of underlying ideologies and values, we are heading for confrontation. Since no movement is homogeneous, both developments may take place side by side. In the end, a community-based standardized but flexible service programme in the form of a computerized service provision will be able to rely on the tamed self-care and self-service attitudes and behaviour of its clientele.

On The Margins of the Information Society: Risks and Benefits for Whom?

I cannot conclude this survey without speaking of those who are likely to appear on the margins of this future society: the new marginal groups. While it is a truism that every society has its marginals, we owe it to those who are likely to be most affected by the incumbent changes to look towards their plight in order to intervene when and wherever possible. Basically, I wish to draw your attention to four groups, three of which can be said to be traditionally marginal while the fourth is potentially universal. You will also find that I refrain from treating the one marginal group that has received great attention in the literature on the future application of microprocessors, namely the handicapped. While I do not wish to denigrate or belittle in any way the very real improvements that modern technology has brought and will bring to alleviate their suffering and compensate for their handicaps, it still seems to me that the purely technical components determining their overall situation in society are relatively minor when compared to the other factors which work for or against their social integration into society.

The Poor Pay More

This most traditional of all marginal groups is likely to suffer disproportionally in an information-based society. The simple reason is, that this future society will depend much more on capital-intensive investment in the home to assure a decent functioning and well-being. Already now, calculations in the energy sector have shown that the poor, here defined as the economically weakest strata, are using almost 10 % more of primary energy for the production of a unit of usable energy, for example, for heating their homes (Joerges, 1979), since they have to use outdated equipment with lower degrees of energy efficiency. This fact has two sides: their inefficient use is in the end more costly to the economy as a whole, but the poor also have to pay more for the same quality of energy than the better equipped households. In the future, the average household will be equipped with a household computer terminal, microprocessor-steered regulation devices for the use of energy, electronic toys and assorted household gadgetry. What at first may still be regarded as luxury, eventually turns into necessity, when other services, still in existence now, will be dispensed with. Most postal administrations in Europe have already plans for the reduction of their customary services, for instance, yet these are indispensable while we are still in transition towards an electronically based mail delivery system and enlarged telecommunication system. One can well imagine the social disparities which are bound to arise from the lack of investment capital among the poor to buy themselves those goods which will be essential for a proper functioning within an information-based society.

Computer Illiterates

The second group I wish to bring to your attention is also well-known, although it periodically changes its appearance, depending on the opportunities and exigencies of the educational system and its links with the economy: the less educated and the occupationally least qualified. The new information technology will have deep effects on the range of skills and competences required both in every day life and in order to ascend the social career ladder. While some, the older members of today's societies, and the least educated, have almost no chance of catching up with the new requirements, the acquisition of new qualifications and skills is not just a matter of generations or transition to a new age. From all studies we know, there is likely to be growing polarization in the qualification structure, the exact extent of which is under dispute (Gizycki, Weiler, 1980) (Council for Science and Society, 1981). While this may mean that the top of the stratificational hierarchy, based on occupational qualifications and the possession of a range of new skills may become more narrow again, with the base becoming proportionately broader, it is another feature which I believe to be potentially more conducive to the constitution of new marginals. I am referring to the decisive shift from the principle of learning for life, still characteristic of the educational system today, towards a system based on the principle of life-long learning. Since it is highly unlikely, for reasons of cost, that it will be the state which can provide a system of life-long education in order to keep up with changing requirements and increasing technical specialization, the burden of cost, which also means the access to opportunities, will be shifted to the individual. Taking into account the highly specialized nature of the competences and skills required, the necessary training will most likely be offered by firms which have a vested interest in screening and offering training only to those whom they have reason to see as most promising in terms of the economic calculations of their returns. While one can envisage a system of commercialized packages being offered for home consumption, the overall picture is not re-assuring. A socially mobile society which is offering restricted opportunities for acquiring the necessary qualifications and skills while shifting the burden of costs and of time to the individual who has to continue his or her learning process during leisure time, is bound to create new marginal groups. They consist of those who are unable or unwilling to take on these additional demands. It has taken centuries to wipe out illiteracy, and to provide all members of society with a modicum of universal education free of charge. What we are facing now is a new and more hideous form of computer illiteracy: more hideous because its demands are higher and more difficult to meet, for reasons of costs, of equality of opportunities and of the individual's capability and willingness to meet these demands.

Women's Place in the Information Society

I have included women as a marginal group, as their relationship to the labour market continues to be tenuous. They are still widely regarded as a reserve labour army, to be relegated to the home according to the expediencies of the market. Despite their impressive inroads into occupational life, women still, and this needs to be repeated again and again, occupy mainly the lower rungs of the occupational hierarchy and everywhere, on average, still earn less for performing the same kinds of jobs than men (Economic Commission for Europe, 1979). In the ongoing discussion on the impact of microprocessors it has been universally acknowledged – to the extent of turning this into a self-fulfilling prophecy – that women, as constituting the majority of office workers, are going to be hit hardest by the job displacement effects. In addition, speculations have been advanced which glorify the home-based computer terminal as the ideal working place for women, since they can at the same time look after their children, help to reduce the costs of communal child care facilities and as an extra bonus, relieve the energy bill of the nation by not needing to use the transport system (Clark, 1981). To relieve women's eventual boredom or isolation (in case she should have time left to become bored) a full array of electronic home gadgetry is put at her disposal which will make household chores a pure delight. In the meantime she can order her shopping via telecommunication and converse with her friends via home-based teleconference systems. While such male fantasies of the combined electronics house-wife and house-worker rightly deserve our ridicule, the underlying structural realities of entrenched male-female relations within the household should make one more cautious. Despite the technological changes within the household so far, detailed studies from several countries show that the ratio division of labour between the sexes within the household has hardly changed (Hartmann, 1981) (Uusitalo, 1981). Time-budget studies covering both Eastern and Western Europe have shown that the ratio of women's work time within the home, compared to that of men, is a disturbing constant (Szalai, 1975). And while the amount of time necessary to carry out certain household chores has been altered as a result of technological devices, the overall time spent on household related work has not changed dramatically. This is mainly due to a change of standards with regard to cleanliness and more time being spent on transport, including shopping and child care (Maimann, Loidl, 1980). This and similar evidence shows that the introduction of technology itself, if the underlying social relations are not touched by it, remains a surface phenomenon. Therefore, women's place in the future information society promises to be as embattled as her place and advances are in today's society.

The Misfits of Information Society

This is the last of the marginal groups and potentially a universal category. Every society produces its misfits – people who, for various reasons, do not fit into the structural slots which have been constructed for them or to which they have been assigned. But while the definition of who becomes a misfit is a very complex social process, the future information society provides a tendentially new definition, which is self-generating. What I mean is the following. Since the advance of modern technology there has been an inherent tendency, as we have seen in the first section, to eliminate the human factor which is regarded as the cause of errors. At first, these efforts were directed at eliminating or at least reducing the human factor in the process of production, mainly by the division of labour and the standardization of tasks to be performed. Concomitantly, there have been efforts directed at the socialization and at the education sector, in order to produce workers who are less amenable to producing the kind of errors that prove disruptive for the production process. These developments are part of the whole process of civilization, as Norbert Elias has charted it for us, with increasing self-control and voluntary subjugation under the norms of behaviour which come with the lengthening of the chains of dependency. On a more concrete level, an unruly labour force has become subject to a self-disciplinary process. With automation gaining control in the factory, a great number of errors have effectively been eliminated and increasing refinement and standardization have greatly reduced the scope for errors. But there is no end in sight. The new information technology guides in very subtle ways the social definition of what passes as error. Such definitions, which are a social process, are often announced as technical necessities, as the famous Sachzwang said, which emanate from a piece of machinery, while in reality technology always allows for more degrees of freedom than the social institutions whose vested interests are at stake, are willing to admit. Information technology also calls, by virtue of the formalized language upon which it is based, for rigorous definitions of standards which serve as inputs, or as outputs which, in turn, generate a more standardized input behaviour. By way of interaction between the technological system and the human actors, standardization of what passes for right and wrong, for reliable and unreliable, is propagated and results in new standards of behaviour. These standards are likely to value highly those qualities that are easily absorbable within the technological system and conform to assumptions about human actors. What I want to emphasize here is that the process of social definition is not a conspiratorial act on the part of a power elite, although it may be used deliberately for such purposes. Rather, it is the result of the interplay of a subtle selection and adaptation process: the existent technology selects those behaviours and standards which are compatible with its operation and eventually other societal institutions adapt in attempting to produce human beings in conformity with them. The reverse side of the process is that of the production of misfits. The crucial difference which information technology brings is that it allows for an almost infinitesimal and objectifiable graduation of adjustments: its universality, its cognitive aspects, its distant storage of information, all combine in allowing refined distinctions as to what an ideal information processing human actor should be like. It will be left to social and political circumstances and contingencies, to draw the line between what passes for ideal, what is still acceptable and who is going to be rejected. The category of misfits is wide open, however.

Options and Conflicts: Towards an Integrative Solution

As we have seen in section three, already a number of tensions have become manifest on the ideological level. Ironically, these tensions arise due to conflicting values, all of which belong to the heritage of Western European societies. If not resolved by institutional accomodation and innovation, they are likely to erupt into conflicts. One kind of tension, inherent in the conflicting values to which the administrative-political and the lay DIY-movements adhere, has already been mentioned. An analogous configuration of opposed values is apparent in the different significance attached to 'own work' as compared to the mandatory DIY-character endemic in a commercialized DIY-economy. Underlying both patterns of conflict is the opposition of the values attached to the autonomy of the individual who is seen as entitled to his or her self-determination in work and care, to the ultimate right of self-realization of his or her human potential, even if this means a regression or lowering of standards compared to professional care or work. On the other side we find the value of ever-increasing efficiency and rationality with the help of a technology, which allows us to minimize errors and to raise output. Associated with the latter are all bureaucratic forms of organization, with industry as its closest ally. A minor ally is the professions which contain still a heterogeneous mixture of an old humanistic residue and a new, socially upward mobile technocratic segment. On the other side of the divide we find groups whose integration into the dominant techno-bureaucratic system has only partially been achieved. These are the young who arrived at a time when employment prospects and, more generally, the outlook on a bright and promising future has been on the wane, and residual groups of a pre-industrial order who have been by-passed by the disciplinary march of industrialization. Finally there are groups who, by virtue of their high educational achievements but declining employment or career prospects that do not match their claims and expectations, are especially prone to seek an alternative life style. There they hope to utilize what they rightly claim to be their knowledge and skills which are apparently not sought after in the established world of careers. Their collective claim for more autonomy and their insistence on alternative forms of socially useful knowledge, of work and mutual care, can also trace its legitimate descent from the tree of the Western value system, just as the branch of bureaucratic rationality and economic efficiency can. But it has to be admitted that the tree is becoming increasingly lop-sided.

Under the worst conditions imaginable – a continuing or even more severe economic crisis which above all affects the welfare state and its achievements in the provision of public services for all and selective additional services and compensatory rights for the extremely disadvantaged – few social conflicts will occur. And when the solution to the crisis is increasingly sought in technological fixes and by ruthless expansion of an economically promising productivity growth while disregarding social costs, severe social conflicts will be inevitable. As with many conflicts of this kind, reaction is likely to be channelled into the defensive rights of those who maintain the status quo of power and privilege, with law and order on their side, while the challengers can easily be labelled as irrational, politically subversive and legally without rights. Under the best imaginable outcome of the worst imaginable conditions, we may end up with a polarized society; its core might consist of those who have well-paid and secure jobs, and who are known to be both loyal and reliable to the values of the dominant political system. The periphery will be vast, consisting of people with highly insecure and lowly paid jobs, who are allowed to fluctuate and keep some of their alternative and autonomous networks, as long as they are not becoming threatening to the dominant order and remain essentially segregated. While this is surely a precarious situation, you may feel free to fill in other variations yourself.

The danger of a polarized society, insidious as the thought may strike us today, is however nothing new in itself. Many societies in history have managed to survive surprisingly long stretches of time with one part clearly dominating the other. An information society, as many observers have also reminded us, has a new range of means of control at its disposal which would in fact make it comparatively easy to ensure the domination of the core over its vast periphery.

But options, which I am convinced we still have, imply that as in a nightmare when one feels the spectre approaching, one still has the chance of waking up. For a society to wake up means, however, that it has to use strategies that are both socially pre-structured and innovative at the same time. The options available are therefore strategies for social integration. These, in themselves, have a double meaning, for it is rare for those who are being integrated, in order to avoid conflict, to be content with the fact and form of their integration. Successful integration always means that an impending social conflict has been averted, that opposing forces and energies have been channelled, if not manipulated, in a manner to which the integrated have given their reluctant consent. Therefore, integration strategies always contain elements of social control, regardless of what other benefits they may eventually bring. In the more recent and more distant past of European societies there have been at least two distinctive instances of the successful practice of social integration strategies. The first one occurred in the midst of the industrial revolution when its ravaging scars threatened to disrupt the societal but above all the economic fabric by the insurgence of the working class. The social security system, institutionalized in instalments in all European societies, served to placate workers' unrest while keeping industrial expansion going. The second instance occurred through the ingenuity of Keynesian economics. It consisted of the successful integration of the social and economic spheres of society which hitherto had been regarded as fundamentally apart. Deficit-spending allowed the economic system to produce that new brand of citizen, the consumer (while the economic foundations for mass production and mass consumption had already been laid by the rise of corporate capitalism). The option that Western societies may well be faced with, through the coming of the information society and the risks and benefits it carries, is that of inventing a new integrative strategy.

The new information technology, while seemingly granting the semblance of autonomy to its users, in fact reduces this autonomy, at least in the eyes of those who aspire to a different set of alternative life styles. While offering the highest professional services on a technological basis, it creates a new dependency of lay persons on experts. With information becoming the all-pervasive commodity, knowledge, especially the private knowledge of the individual, is further reduced in its utility and made obsolete. The promises and potential to 'personalize' communication and provide a range of services to suit the whims of the individual consumer militates against, and is in contradiction with, the social imperative to build new collective identities. In the end, the kind of hyper-individualism which is promoted by the foreseeable uses of the new technology, is socially self-destructive. The integrative strategies which may be needed in order to face the challenges of the information society, will have to take these and other contradictions into account, which the new technology serves to bring to the fore.

Reference

1   I am indebted to Manfred E. A. Schmutzer for pointing this out to me.

Bibliography

Rosenberg, Nathan, 'Technology and economic growth' in: Cross, Nigel. Elliott, David and Roy, Tobin (eds), Man-made Futures (Hutchinson & Co., London, 1974)

Noble, David, 'America by Design' (Alfred Knopf, New York, 1979)

Rada, J., 'Microelectronics, Information Technology and its Effects on Developing Countries' in: Berting J., Mills S. C. and Wintersberger H. (eds), The Socio-Economic Impact of Microelectronics (Pergamon Press, Oxford, 1980)

Rada, J., 'The impact of micro-electronics', ILO Geneva (1980)

Norman, Colin, 'Microelectronics and the World Economy', Transatlantic Perspectives 4 (January 1981) 19–22

Braverman, Harry, 'Labor and Monopoly Capital, The Degradation of Work in the Twentieth Century', Monthly Review (New York, 1974)

Gorz, André, 'Adieux au prolétariat', Editions Galilée (Paris, 1980)

Benedetti, F. de, 'The impact of electronic technology in the office', Financial Times Conference, Tomorrow in World Electronics (London, March 21–22 1979)

The Engineer (September 14, 1978) 24–25

Salomon, Jean-Jacques, 'Adhésion ou résistance au changement technique?' FAST Project, Bruxelles, mimeo (1981)

Elias, Norbert, 'Über den Prozeß der Zivilisation', 2 Bd., (Suhrkamp, Frankfurt a. M., 1976)

Kalbhen, Uwe, Krückeberg, Fritz, Reese Jürgen (Hrsg.) 'Gesellschaftliche Auswirkungen der Informationstechnologie' (Campus, Frankfurt a. M., 1980)

Gershuny, J., 'After industrial society. The emerging self-service economy' (McMillan, London, 1978)

Skolka, J., 'Long-term effects of unbalanced labour productivity growth: in the way to a self-service economy' in: Solari, L. and Pasquier, J. N. du (eds) Private and enlarged consumption (North-Holland, 1976)

Joerges, Bernward, 'Berufsarbeit, Konsumarbeit, Freizeit', Soziale Welt, Jg. 32 (1981) Heft 2, 168–195

Maimann, Helene and Loidl, Josef, 'Technischer Wandel – sozialer Wandel: Das Beispiel Haushalt' Interdisziplinäres Forschungszentrum Naturwissenschaft-Technik-Gesellschaft (Wien, 1980)

Giedion, Siegfried, 'Mechanization takes Command' (Oxford University Press, New York, 1948)

Jansehen, Doris, 'Rationalisierung im Alltag der Industriegesellschaft' (Campus Verlag, Frankfurt a. M., 1980)

Jansehen, Doris, 'Kommunikationstechnik im Alltag', Wissenschaftszentrum Berlin, mimeo

Reese, J., Kubicek, H., Lange, B. P., Lutterbeck, B. and Reese U., ,Gefahren der informationstechnischen Entwicklung' (Campus Verlag, Frankfurt a. M., 1979)

Sola Pool, Ithiel de, 'Technology and Change in Modern Communication', Technology Review (November-December 1980) 65–75

Actes du Colloque International Informatique et Société, Informatique Télématique et vie quotidienne, vol. III. Série Impact

Cowen, Robert, 'New Literacy for the Computer Age', Technology Review (May-June 1981) 8

Lange, Bernd-Peter, Kubicek, Herbert, Reese, Jürgen and Reese, Uwe, ,Soziale Informationstechnologie als Programm', Gesellschaft für Mathematik und Datenverarbeitung (Selbstverlag GMD, Bonn, 1980)

European Centre for Social Welfare Training and Research, Expert Meeting, ,Use and Abuse of Social Services and Benefits', Montreux (10–14 November, 1980)

Joerges, Bernward, 'Die Armen zahlen mehr – auch für Energie', Zeitschrift für Verbraucherpolitik 3, 2 (1979) 155–165

Gizycki, Rainald v. and Weiler, Uwe, 'Mikroprozessoren und Bildungswesen' (R. Oldenbourg Verlag, München–Wien, 1980)

Council for Science and Society, 'New Technology: Society, Employment and Skill' (The Council for Science and Society, London, 1981)

Economic Commission for Europe, 'The Economic Role of Women in the ECE Region' (Geneva, 1979)

Clark, John, 'Automatic Housewife: who are the future casualties of a computer invasion of the home?', Perception (March–April 1981) 14–15

Brunet, Lucie, 'A la merci des machines misogynes', Perception (March–April 1981) 21–24

Hartmann, Heidi, 'The Family as the Locus of Gender, Class and Political Struggle: The Example of Housework', Signs, vol. 6, no. 3 (Spring 1981) 366–394

Uusitalo, Liisa, 'Differentiation of the way of life by technological development, social system and sex', University of Helsinki, Research Reports, no. 9 (1981)

Szalai, A., 'The Situation of Women in the Light of Contemporary Time – Budget Research', Conference Background Paper to the UN World Conference of the International Women's Year, Mexico City (1975)

Not Quite Human: Science and Utopia

The Non-existent Science of Utopicists

While preparing for this contribution I went to see a film: Sans Soleil by Chris Marker. In 100 minutes a dense collage of visual poetry is presented to the spectator, accompanied by an equally dense essay of impressions collected in Japan and Africa. Japan has been chosen as one possible society of the future, representing what the film pictured to be one extreme in the art of survival of a civilization yet to come. What fascinated me was the utopian touch that was carefully and yet emphatically, read out of the present: the music of video-games, for instance, as the constant, underlying musical theme of a buzzing metropolis; a description of how these games were programmed and how a new collective language of imageries was in the making, coding memories and thus providing the essence of a future collective unconscious. Interspersed with everyday scenes, celebrating their banality and uniqueness at the same time, the film cautiously proceeded to construct an imagery of future, in which humankind continues to evolve, guided by the computer and compuotional thinking. The emphasis was put on the collective mind, and not the individual, in the making, and how this new form of technology-based consciousness would interact, shape and be shaped by what the film-maker sought to single out. Japanese society was predisposed, in his view, to serve as a model for survival, because it knew how to balance high technology with the mechanism essential for survival – social ritual. Whether these involved prayers for animals or for the spirit of material things, ceremonies of purification, of expressing joy or channelling aggression, mind and – the social – body were pictured as meeting in a gracious, convincing and yet for a Westerner deeply irritating way. The film made no concession to the Western image of Japan; no allusion to the race for the fifth generation of intelligent computers or to the hot economic climate of intense competition appeared, nor any of these themes that figure prominently in the current Western debate – control of data banks and fear of more comprehensive surveillance through state or large corporations; intellectual property rights and the issue of secrecy; the possible isolating effects of the new technologies when substituting standardized expert systems, the artificial experts, for conversations with human experts or with friends. And yet, in its non-intentionality, its obvious digression from the dichotomous mode in which future developments are often

Reprinted with permission from Trappl R. (ed.), Power, Autonomy and Utopia, Plenum Press, New York 1986.

presented in the West, the film offered a much more convincing image of what one possible future in the mind and computer age might look like, than any other account I have come across.

Societal imageries of possible futures are not a thing of the past, although the grand visions of entire societies to be built have apparently given way to a much more fragmented view, either built around minority groups in society or split into a myriad of individualized micro-utopias (for more details, see Mendelsohn and Nowotny, 1984). In running briefly through the history of utopian thought, I would like to focus on the inherent tension between science and utopia, as well as on some of their commonalitites. Next, I will turn to the field of AI and robotics, as an interesting example of what actual developments in this rapidly evolving field of research and applications mean for the utopian-dystopian scheme of projecting future developments. Finally, the question of the tension between science and utopia will be taken up again and I will ask whether we are not witnessing the emergence of a new utopia – one that can tell us, perhaps, more about the present than about the future, but which will not fail to influence what the future will look like.

The still fermenting field of AI, cybernetics, systems theory and their application – as also the occasion for this scientific gathering demonstrates (Trappl, 1984) – has perhaps as no other recent field of growth of scientific knowledge and engineering, both, such a long history in utopian and dystopian thought and actual developments that appear to have the potential of realizing what has been anticipated in a negative and positive version. It offers an experimental workshop in utopian-dystopian thinking, inviting comparisons, for instance, of the old and recurrent themes that feature robots or other artificial human-like constructs or thinking machines with what has already been realized or is in making (Fleck, 1984). One could re-analyze predictions that have been made in the past, sinister warnings as well as blissful prophecies regarding a cornucopeian future, and point to their inaccuracies, their faults in reasoning and their failure to grasp essential constants – but such an exercise, unless much more fully developed, would not necessarily guard us against committing similar errors today.

In fact, it was one of the premature and yet audacious ideas put forward by Otto Neurath, here in Vienna some 65 years ago, to work towards the science of utopistics (Neurath, 1979). What he meant was a kind of early technology assessment, with the crucial difference that it was not an isolated technological system or a singular technological development that was to be assessed with regard to its likely future consequences for society. Rather, utopian systems were to be systematically compared with each other, in order to detect the flaws in their reasoning or in their methods of extrapolation. They were to be tested in the usual scientific way, but in a kind of rigorously controlled thought experiment. For Otto Neurath, the great visionary of a new social and scientific order, who conceived

84

of himself as a social engineer in the most noble connotation of this word, any utopistic scheme meant planning for the rational basis of societal life, according equal importance to our knowledge of its social, and of its scientific-technological foundations, a programme unachieved until this very day.

Utopian and Dystopian Imageries of the Past

In the absence of the development of such a science of utopistics, the imaginary constructions of ideal societies, including the place accorded in them to science and/or technology, or of parts thereof, can be analyzed in a historical mode (Elzinga and Jamison, 1984). The beginnings of the utopian imagination in Western Europe, were still modelled after the religious world-view of the times; it were spiritual and religious ideas that served as guide-posts for the more mundane programmes of how to construct ideal societies on earth, while it was only with the advent of modern science and technology that secularization set in here as well (Manuel and Manuel, 1979). Technology, then as now, seemed to offer an easy way out of otherwise intricate social problems: it promised the fulfillment of material wants for all, beginning with Francis Bacon, and an end to human misery; aggression and wars would become superfluous and even the daily disruptions and irritations would be eased away by technological efficiency (Bacon, 1627).

It was left to the ideologically responsive function of science to take up the promise of consensus and harmony for a society full of internal strife and disorder, as was the case with England in the 17th century. Utopian thought captured the imagination everywhere as an endeavour to keep disorder at bay. In this function, it inevitably became loaded with a surplus of order, both in a positive and negative sense, that it was not able to shed ever since. In an age, in which turmoil and incoherence of actual social life were palpably felt, utopian writings were idealizations directed towards organizational and bureaucratic order. They prefigured either liberal or authoritarian tendencies that led eventually to the rise of the absolutist and the modern democratic nation-state. But once science had made accessible the "marvellous symmetry of the universe", its orienting function was to transform rationality and celestial harmony into the guiding vision for the architectural social structure to be implemented in this world: the cosmic prepetuum mobile became the model for a social utopia which, once set into motion, was thought to function perfectly forever, if and when similar universal laws, applicable to human behaviour were found (Winter, 1984). Utopia definitely ceased to be a Christian-inspired heavenly Jerusalem and became a state which could be brought about through action, guided by science, while connecting the idea of scientific feasibility with universal happiness. The modern emerging enterprise was quickly turned into a rational as well as a utopian vehicle, charged to bring about a social world constructed in its mirror image. What else could pose as the unsurpassed master copy for a social order to be built than the natural order with its display of invariance, harmony and eternal laws?

But neither the permanent technological fix, nor the celestial harmony inspired by scientific discoveries could in the end bring about the realization of utopia NOW. The utopian horizon kept moving onwards, not least because of the progress achieved by science and technology. In the middle of the 18th century, the classical space utopia, governed by rational, geometrical constructions in which the interests of the subjects were held to be congruent with those of the social commonwealth, gave way to the dynamic time utopia, in which a more open construction prevailed, reflecting also a change in the conceptualization of time (Luhmann, 1980; Koselleck, 1979; Nowotny, 1975). The role that science and technology played in these transformations is not simply one of empirical inductions. Rather, science and technology created a horizon for the myth of the history of reason to unfold. The actual progress achieved provided the empirical substance of verifiable experience, on the basis of which the projection of the hypothetically possible occurred. Science and technology provided the methods, content and ideology to make a certain kind of future thinkable.

It was a future deeply molded by the belief in progress. It became a general rule for scientific and technological inventions to lead onto new inventions which to predict precisely in advance was not possible, but which gave new space for the utopian imagination as well as providing the verifiable background for the belief in progress. Progress became the dynamically stabilized difference between experience and expectation – not yet tarnished by the shadows of its more negative side-effects (Koselleck, 1979).

One of the most salient characteristics of the scientific and technological optimism which radiates throughout the 19th century is its seeming smoothness. It expressed itself equally forceful in utopian writings of the time. It was achieved, not through the miraculous workings of celestial harmony applied to fragile social constructions, nor through the ingenious congruence of individual and collective desires and actions, so characteristics of earlier utopias, but exclusively through the command that science and technology offered in reshaping social relations. Smooth functioning precludes, among other things, ordinary discontinuities as well as major catastrophes – an interesting gap already contained in Bacon's New Atlantis. In it, catastrophes are prevented from occurring in nature, because they have been subject to human control. Bacon, the fallen statesman, who knew well from his own life what catastrophes meant, eliminated them altogether from his vision of the future. Other utopias followed in the same vein. For the present generation, it is hard to imagine how arduous the belief in social happiness was that became the hallmark of the social utopias of the 19th century and the role played by science and technology in this scheme. Smoothness in operation as the guarantee for happiness, followed by a future built upon industrial work and, not surprisingly, social order was projected as functioning as easily as a well-run production plant. The direct line of descent of this notion can be traced right through to the enthusiasm with which the ideal of social planning was to be received in the early part of this century.

It was left to the rising dystopian vision, the correcting device for the excessive zeal of the utopian imagination, which would henceforth and irreversibly split the social order into those who controlled and those who were controlled, to bring into the open the underlying tension between utopia, conceived as an ideal societal construction, and science. As J. C. Davis has argued at great length, an inherent dilemma remains between utopian thought and scientific development, as long as science has the endless capacity for innovation and hence, for altering the conditions of social life as well. Utopia, in its desire to control and impose an ideal order, cannot tolerate in the end that which is fortuitous, spontaneous and which threatens to undermine its carefully constructed laws and ideals (Davis, 1984). Brought out in a bitterly satirizing or grossly exaggerating way in many dystopian examples, the social order is depicted as controlling every innovative, original, spontaneous act or thought since it threatens to undermine the order already established. Science, like falling in love, is accorded in this construct the subversive and dangerous potential for evading or circumventing established laws of social thought and conduct, by allowing the unexpected to happen. For this is the other side of the coin: while utopia and science share strong tendencies to reduce contingencies to laws and to build upon invariants, science alone, according to Davis, is an open-ended dynamic process, in which the unexpected, serendipidous and the accidental can still occur and are highly valued. The utopia imagination, Davis maintains, cannot possibly match the multitude of possibilities offered by science. The kinetic – moving – utopia is therefore a myth. It expects utopia to predict the course of future scientific innovation which, however, remains unpredictable in its core. But before proclaiming that utopia will either stop science or be overthrown by it, I suggest to examine what the rapid growth of AI and its applications have meant so far for utopian and dystopian thought and what new and unexpected twists the present argument might take.

AI: An 'Automatic' End to Utopian Thought?

(Note: this is the title with which James Fleck (1984) nicely captures the twist of the argument).

Ideas of artificial human beings or thinking machines have pervaded legend and literature from the earliest times. But it is only in the last 20 years or so, that technologies such as AI or industrial robots have appeared which seem to have the potential to realize these ideas. When formerly magical knowledge was seen as capable to produce artificial, humanlike constructs, they stand today as symbols for scientific and technological advances in general. Yet, as Fleck shows, some of the underlying themes that kindle the utopian-dystopian imagination have remained surprisingly constant: the themes of robots as dangerous knowledge, and robots as projections of Man, of men and women, and of what being human is essentially all about.

It would lead too far to retrace here in detail the recurrent permutations of a few overriding themes that are to be found in the pertinent utopian-dystopian literature and above all in science fiction, the newly specialized branch of its more general literary predecessor. Fears and hopes seem to be triggered almost synchronically by the same developments, although dystopias are clearly gaining ground. When surveying the contradictory images that are thus created, one is struck by two observations: first, by the fact that the dominant themes and images, arguments and apparent refutations, are by no means confined to the literary domain alone, but are equally strong characteristics of the ongoing critical discussion on the social impact of the new technolgies (Bjorn-Andersen et al., 1982). Moreover, the utopian-dystopian line runs through the camp of practitioners as well as through the camp of their critics (one of the most prominent and early critics among the ranks of practitioners was Joseph Weizenbaum). While in the science fiction literature, for example, it is the survival of the human race, which is at stake, threatened to be overtaken by the artificial constructs that resemble them to perfection while topping them in efficiency and achievements, in the political discussion it is the extent to which machines will replace the human work force. In the literary genre, the difference which separates true humans from their imitations has been treated in many permutations, emphasizing what is thought to be specifically human. It compares well with the ongoing debate on the possibly dehumanizing effects, once expert systems will be widely used, which again centers on what is thought to be the essence of human communications and interaction. Parallel warnings, for instance, are also raised recurrently with regard to the dangers of centralized control which technologically sophisticated systems facilitate, but also touch issues such as the preservation of cultural variety (Negrotti, 1984).

The other observation pertains to the utopian-dystopian dichotomy, so characteristic again of both, the literary-fictional and the actual, political discourse. It seems as though future developments function as an immense screen for the projection of present social interests and for the extrapolation of present hopes and fears. Exhortations and warnings oppose each other in rhetorics and argumentations, which can easily lead towards an eventually sterile debate. In summarizing attitudes towards thinking machines on the part of AI practitioners and outside critics, Fleck distinguishes between a simple utopian ideology of AI; the simple dystopian view of AI, which consists essentially in asserting its reductionist nature; and two more differentiated positions: one asserting that AI may be dehumanizing because it embodies an alien technological rationality, while the other view proposes that AI offers a way out, i. e. of humanizing technology, because it takes explicit account of human cognition. While this is a familiar controversy by now, its argumentative structure is by no means limited to AI alone.

Underlying this dichotomous mode of reasoning is a deeply embedded tension, which is inherent in the nature of the technology developed in Western societies over the past 200 years, in consort with the social values that support it. As Langdon Winner has pointed out, this technology has always been most productive, when its ultimate range of results was neither forseen, nor controlled (Winner, 1977). It always does more than intended. While this has been regarded in general as a welcome feature of technology up to now, since it serves as basis for the next round of ongoing developments, the unintended consequences are increasingly becoming more visible and questioned. The unstated common assumption until now was, that the positive consequences would – automatically? – outweigh the negative unintended consequences. It is this unstated assumption, which seems no longer valid. We have to face the fact, finally, that in the forces that gave rise to the development of science and technology, unintended consequences were not *not* intended (Winner, 1977).

Looking back to the early enthusiasm that AI and similar developments engendered, to "the days when everything seemed possible", and when one of the brightest dreams was the creation of a programme that would mimic all human problem-solving abilities, we can clearly see that little thought was accorded to exploring the full range of second-order consequences. Even if today's assessments of the actual achievements, as Mitchell Waldrop puts it, both in the engineering camp, who are trying to get their programmes to do smart things, and in the scientists' camp, who are after a general theory of intelligence, are much more modest. Waldrop maintains that the main thing that AI researchers have gained on the theoretical front is a certain humility, and of how much a computer has to know before it can do much of anything (Waldrop, 1984) but the vision of the revolutionary potential lingers on, both among the practitioners and among the general public, irrevocably interwoven with dystopian elements.

A New Utopia in the Making?

In the film I mentioned in the beginning, one of the dimensions lending credibility to the potential for survival of the Japanese society in an age dominated by microelectronics, computer technology and computational thinking, was the persistence of social rituals. Like all rituals they serve to symbolize relationships, including those that connect human beings with "the spirit of things". It is one of the paradoxical and unanticipated consequences of the development that science and technology have taken, one of the evolutionary turns in the conceptual apparatus of societies, that this archaic notion of communicating with things, which modern science from the 17th century onwards has declared to be devoid of spirit and to be nothing but dead matter that can be controlled by the human mind, takes on new meaning and relevance in the complicated relationship between human beings and intelligent machines. By fusing 'mere matter' with intelligence, by simulating, imitating, and by partly perfectioning the functioning of human reason, of thought and language operations, through such mediums as dedicated (!), massively parallel machines or of intelligent knowledge-based system architecture, an important shift is taking place. While previous scientific discoveries touched upon and redefined the place of humans in the natural order, the present cultural environment has since long replaced it through its own artefacts. The crucial relation now becomes that of humans to their own creations – to their material products.

One way of re-defining this relationship consists in the exploration of the basis of consciousness, including consciousness that resides in and can be discovered as well as imputed into 'mere matter' and artificial things. AI, in the eyes of many of its practitioners, contains a revolutionary potential which is based on the vision of a new epistemological approach, as yet only dimly understood.

From a relatively simple set of ideas and concepts, reflecting the prevailing dichotomous epistemology, work in and around this area is beginning to generate an enriched vocabulary and an artificial interpretative structure. This extended cognitive space, as it percolates into the wider ideological structure presently dominated by the ditochomous epistemology of man vs. machine, mind vs. matter, intuition vs. calculation, and subjective vs. objective, will ensure that the advent of robots and AI, no matter how limited or spectacular their capabilities, will be absorbed without either of the major simple utopian or dystopian outcomes being realized (Fleck, 1984). This new approach, which can be interpreted as the basis for nothing else but a new utopia, leads to the reintroduction of mind, albeit on a material basis. "It is believed that the new approach can accomodate contingency, chance and individual variability, without any intent to eliminate them. By challenging the man/machine and subjective/objective dichotomies, what is sought is not the extension of natural law to cover man, but rather, the elimination of a purely instrumental conception of science and the re-introduction of mind, albeit on a material basis, into the operation of the material world" (Fleck, 1984).

In one of her perceptive essays on the impacts of AI, Margret Boden examines progress in AI, both in core research areas that are likely to make rapid progress within the next decade and in what she calls the programme of long-range AI research (Boden, 1984). In observing impacts of AI developments she notices foremost that AI will influence other sciences in their general philosophical approach as well as in their specific theoretical content. In her opinion, psychology and to a lesser degree biology have already been affected by computational ideas. While the behaviourists in particular had outlawed reference to mind and mental processes as unscientific and mystifying, AI, based as it is on the concept of representation, has rendered these concepts theoretically respectable again. Moreover, one should add, it has opened up a new and rapidly expanding field, called the cognitive sciences, and has led to the first, tentative formulation of what is called the Cognitive Paradigm (de Mey, 1982).

But the new relation to "the spirit of things" or to the mind embedded in matter is not only a visionary programme for what is perhaps a new (utopian?) epistemology in the making. It is also to be found in social practice, here and now. Sherry Turkle has given a fascinating account of what she calls the "subjective computer" – the use of personal computers and the highly emotionally charged atmosphere in which users are working out their feelings of power and control, of being safe in a protected environment (Turkle, 1982). She suggests that the computer serves largely as a projective screen for other personal concerns. By many people it is experienced as an object betwixt and between, hard to classify and hard to pin down. She describes in the words of users how the elusiveness of computational processes, the tension between local simplicity and global complexity is experienced and contributes to making the computer an object of projective processes. In view of the computer's internal processes, individuals project their models of mind and in the descriptions given of the computer's powers, people express feelings about their own intellectual, social and political power – or their lack of it.

Thus, it is not surprising to encounter again some of the oldest anthropomorphic imageries, but also a yearning for security, for the possession of a safe corner of reality, amid another outside reality which offers it only to a small degree. The users described by Turkle are far from having an instrumental relationship with their computer, nor are they playful in the narrow sense of the word. Rather, they are very serious in wanting their computers to have a transparency that other things in their life do not have. The social world and the world created by science and technology seem to complement each other once more: the utopian pendulum can be observed in motion. And while it is easy to relapse into the utopian-dystopian mode of thought by interpreting the potential of the new relationships with personal computers as humanistic and hence beneficial, or by condemning them, by insinuating that, once they are widespread, they may become the new opiate of the masses, we should instead return to the embodiment of the new utopian pendulum, observing its swing between science and utopia a bit closer.

The Utopian Pendulum in Motion

The mutual attraction and threat which science and utopia pose to each other in their common desire to subdue the contingent, has been described by J. C. Davis as the two horns of a dilemma (Davis, 1984): utopia can cope with science only – since science will inevitably change existing social arrangements and therefore threatens to destabilize them – when it conceives of a society that allows its members to control the moral and social consequences of scientific and technological discovery. This is a familiar dimension in utopian writings from the 18th century onwards until the present debates. Within such a utopian construct, the temptation is great to attribute fixity to science. In its extreme, the accidental aspect in scientific discovery would have to be removed, the spontaneous discovery harnessed in advance. Only then would it no longer menace the stability of the preconceived perfect social order, only then, presumably, would it be possible to extract only the beneficial yields of science and technology, while suppressing the negative ones.

The other side of the dilemma is the following: if science is not to be completely controlled and thus being reduced ultimately to a static and closed system, the ideal society has to be conceived as changing in a dynamic, evolutionary way. But can the utopian imagination really conceive of a continuous and endless sequence of legal, institutional and adminstrative devices, Davis asks, not only capable of adapting to successive changes, but also capable of guaranteeing their own transformation? Davis' answer is a clear no. Utopia will either stop science or be overthrown by it.

If it is impossible to foresee and to control all future consequences, intended and not intended ones, positive and negative ones, that will result from ongoing scientific and technological work, does it mean that a rampant technology has to be accepted? Put differently, are we stuck in the endless and sterile debates in the utopian-dystopian mode, until actual developments overtake the limits of the imagination by producing a much more differentiated pattern? For utopias and dystopias are always mirror-images of the societies that produce them; they are collective representations of the hopes and fears that these societies harbour with regard to a future that does not yet exist. Since utopian and dystopian thought are temporarily rooted in the present, they also tell us more about the present than about the actual future. In reading them as expressing the present oriented towards a future, and by observing and analyzing actual developments in their deviations from what has been hoped or feared, we are led eventually to a better understanding of how the future is actually made today.

For an observer of the contemporary scene, the future, once dreamt about in a paradisical or nightmarish way, has come to stay. While it is easy to be overly impressed by the scientific and technological forecasts that have been realized and have actually provided the islands with plenty and wishfulfilment, at least for that part of humanity that lives in the rich industrialized nations, its dark side has also come to stay with us. We have seen how the utopian-dystopian tension has moved along with the continuing debate about the social impact of science and technology's latest achievements, but we have not sufficiently appreciated the interaction between the social side of this development and the technological one. The apparent inability to synchronize rates of change, to adapt them to each other, to humanize technology and to invent new social rituals that will allow social beings to come to better terms with their own artificially created products, is the hidden message of the utopian-dystopian accounts. Whenever social problems are pressing, redemption is sought on the side of science and technology. Whenever their impact is perceived as potentially de-stabilizing, de-humanizing and threatening the social fabric, visions of a new society are created and their dystopian mirror-image signals an impending catastrophe.

Looking backwards, it is rather obvious that neither have science and technology been stopped by utopia, nor has science victoriously swept aside all utopian thinking. Quite on the contrary, utopia and dystopia have entered science and are here to stay. While it is impossible for the utopian imagination to anticipate or even keep pace with the actual developments of research and innovations from the outside, it has come to orient these developments from the inside. In doing so, – and discussions on AI and its impact are a good illustration – the utopian-dystopian tension is partly continued, but has partly been superseded by a new utopia: how to reconcile matter and mind, how to find the key to a new understanding of the universe in exploring the secrets of consciousness. The incorporation of utopia means also that the present becomes more and more loaded with choices. While science is seemingly producing a multitude of possible futures for our disposal, there can still be only one present. The hot fields in which present scientific utopias are taking material shape, show how a possible future is reduced to an instant present. The radical consequence to be drawn from this merger of science and utopia today is perhaps to realize that we are contributing ourselves to utopia and dystopia in the making and are confronted with having to live with them at the same time.

Ernst Bloch, one of the great writers on utopia and a utopian himself, wrote of the final stage: "Es soll zu guter letzt, wenn keine Utopie mehr nötig ist, Sein wie Utopie sein" (in the end, when utopia is no longer necessary, to Be shall be like Utopia). Perhaps science has brought us closer than we ever imagined we would come, to the obligation of reconciling actual Being – the social side – with Utopia – the scientific and technological side.

Bibliography

Bacon, F., 1627, New Atlantis.

Bjorn-Andersen, N., Earl, M., Holst, O., and Mumford, E., 1982, "Information Society: for richer, for poorer", North-Holland, Amsterdam.

Boden, M., 1984, Impacts of Artificial Intelligence, Futures, 2:60–70.

Davis, J. C., 1984, Science and Utopia: The History of a Dilemma, in: E. Mendelsohn and H. Nowotny, eds.

Elzinga, A., and Jamison, A., 1984, Making Dreams Come True – An Esay on the Role of the Practical Utopias in Science, in: E. Mendelsohn and H. Nowotny, eds.

Fleck, J., 1984, Artificial Intelligence and Industrial Robots: An Automatic End for Utopian Thought?, in: E. Mendelsohn and H. Nowotny, eds.

Koselleck, R., 1979, "Vergangene Zukunft", Suhrkamp, Frankfurt.

Luhmann, N., 1980, "Gesellschaftsstruktur und Semantik", Suhrkamp, Frankfurt.

Manuel, F. E., and Manuel, F. P., 1979, "Utopian Thought in the Western World", Havard University Press, Cambridge, Mass.

Mendelsohn, E., and Nowotny, H., eds., 1984, "Nineteen Eighty-four: Science between Utopia and Dystopia (Yearbook in the Sociology of the Sciences, Vol. 8)", D. Reidel, Dordrecht.

de Mey, M., 1982, "The Cognitive Paradigm, Sociology of the Sciences Monographs", D. Reidel, Dordrecht.

Negrotti, M., 1984, Cultural Dynamics in the Diffusion of Informatics, Futures, 2:34–46.

Neurath, O., 1979, Die Utopie als gesellschaftstechnische Konstruktion, in: "Otto Neurath, Wissenschaftliche Weltauffassung und logischer Empirismus", R. Hegselmann, ed., Suhrkamp, Frankfurt.

Trappl, R., ed., 1984, "Cybernetics and Systems Research 2. Proceedings of the Seventh European Meeting on Cybernetics and Systems Research", North-Holland, Amsterdam.

Turkle, S., 1982, The Subjective Computer: A Study in the Psychology of Personal Computation, Social Studies of Science, 12:173–206.

Waldrop, M. M., 1984, The Necessity of Knowledge, Science, 223:1279–1282.

Winner, L., 1977, "Autonomous Technology", MIT-Press, Cambridge, Mass.

Winter, M., 1984, The Explosion of the Circle: Science and Negative Utopia, in: E. Mendelsohn and H. Nowotny, eds.

Does It Only Need Good Men to Do Good Science? (Scientific Openness as Individual Responsibility)

From Bacon to Bernal: Does It Only Need Good Men to Do Good Science?

Ever since its inception modern science has continued to adhere to one guiding utopian vision: that of the ideal scientific community. Faced with the 'greatest public unhappiness' and 'while the considerations of Men and humane affairs may affect us with a thousand various disquiets', as Thomas Sprat put it in 1667, the scientific community offered 'room to differ, without animosity' and the enquiry into nature would even 'permit us, to raise contrary imaginations upon it, without any danger of Civil War'.[1] It would not only offer new procedures for settling differences by argument and experiment, but the Court of Reason would eventually become the place in which all matters, including the unhappy, public affairs would be adjudicated.

In its essence, it is a community composed of an elite that – paraphrasing Bernal[2] – I will call good men doing good science. The good men are those who have espoused scientific rationality as the guiding 'scientific world view', as Otto Neurath was to call it in the 1930s in the wake of a new enthusiasm for planning based upon science.[3] The good men were in possession not only of the weapons against darkness and ignorance which every age seems to bring forth in a new guise; they were also the carriers of an explicitly formulated claim towards a kind of moral superiority based upon the inherent intellectual superiority of scientific rationality. Contrary to science's official apolitical stance, the better insights that science provided were not limited to the scientific realm properly speaking. Rather, what the utopian programme and its claim offered was their extension into the unruly realm of human affairs. If only scientific rationality were to reign and guide their management, thus runs the deep conviction of the good men doing good science, order would at last be brought into the messiness of social life.[4] The final test would be 'this great enterprise of our time, testing whether men can . . . live without war as the great arbiter of history',[5] ie whether science would in the end be able to supersede the recourse to physical violence as a means of settling differences; and whether, on a less existential scale, a

Reprinted with permission from Gibbons M. and Wittrock B. (eds.), Science as a commodity, Longman 1985.

more efficient management of human and natural resources would supersede the wastefulness that now prevails in social and economic relations.

The claim of goodness is therefore a double one: it needs good men to do good science – meaning that in science only the best will achieve – but also that good men doing good science are good at something else: they extend the boundaries of science in the world, the process of seemingly limitless expansion in which science is engaged. In the imagery of the conqueror, the liberating and progressive consequences of this expansion clear the way for a more emancipated humanity in general. This theme of the scientific elite being the best and only guarantee of good science and the good state of human affairs has been relatively unchanged from Bacon to Bernal.

Yet, even when looked upon as a utopian vision the deficiencies in its claim towards moral superiority are apparent: Bacon's *New Atlantis*, grandiose as it was as a prophetic scheme, remained a fragment. The other part that Bacon intended to write on the Best Imaginable State, was never written. We may also note that Bacon who, as a statesman, knew what catastrophies meant and how they occurred, has carefully eliminated them altogether from his island of scientific utopia. What power and knowledge, after having met in one, would ultimately be used for, remained a question left for his intellectual descendants to answer. Likewise, Bernal has eliminated any traces of incoherence or disorder from his vision of the social function of science. He had the greatness, as Ravetz has rightly reminded us, to perceive clearly how applications of science can be blocked or distorted by commercial greed and to analyze how the cycle that leads to and from human needs towards the application of scientific research can be interrupted, distorted or destroyed by secular institutions. Yet, he was also firm and convinced about the simple solution he put forward: change the social order, which is now unjust, inefficient, downgraded by the vices of rampant capitalism, and you will be able to change science.[6] Today, with the sad benefit of hindsight, we know that this simple change in context of application has not eliminated the abuse of power; rereading the writings of those times we are appalled by the lack of sensitivity to our present predicament of what can go wrong and by the strong scientistic and technocratic elements in what a 'world adapted to man' looks like. For historical experience was to teach us that there was no end to the new problems arising after science had tipped the economies of scale and began to move towards becoming more fully industrialized, bureaucratized and militarized.

To speak about the individual scholar and the scientific community and, within this relationship, about openness as the responsibility of the individual is insufficient. It is simply not enough to look at the human, individual frailties without having a sufficiently clear grasp and understanding of the imperfections of institutions that produce and shape individuals. It will not be possible to change the men who do science without taking into account how science has changed in producing certain types of men (I will revert to the question of women doing science later). Science has extended into so many other areas of life, but, as other conquerers have had to experience, it has assimilated certain traits of the conquered in doing so. The claim to moral superiority and to leadership in human affairs has been severely battered. With inevitable clarity, we know today that things can go wrong: good men can be drawn into doing bad science.

The Organization of Complexity: The Case of Molecular Biology

The organizational model, upon which the premise of good science as the convergence of high standards in moral terms and scientific work alike was based, has undergone a profound transformation. The university, once the traditional home for the systematic production of knowledge and the transmission of learning, has had to make room for its former adjunct, the research laboratory, which has taken on an organizational life of its own. Recent studies in the sociology of science have prided themselves on discovering 'life in the lab' as the 'factory in which order is produced'.[7] As in other factories in which capital investment is taking place, the operation is geared towards production – in this case, of scientific facts which have to be stabilized, ie standardized, and to be invested with credibility in order to be traded, sold, appropriated. The scientists working in this factory are said to be obsessed to a certain degree with the economic categories of success: they speak in terms of the costs they incur, they think in terms of cost-benefit analyses, they bargain with each other in the 'manufacture of knowledge' and they behave like bankers who control budget and balance accounts, payoffs and tradeoffs alike. The rules that dominate the research game are in their turn dominated by the economic laws of investment and return, of profitability and success.[8]

Yet, comparing these studies with those that have come out of industrial sociology, we may wonder why it has taken so long to discover a certain convergence in the games that scientists play in the laboratory and the rules that govern behaviour on the industrial shopfloor. Behind such an analogy lies a process of convergence of new principles of organizing complexity in science and in society alike. As an illustrative example of how such convergence began to take hold, consider the role played by the Rockefeller Foundation in the 1930s in the establishment of a new field that would eventually become molecular biology.[9] A new network of informal personal contacts in an as yet unestablished field was created among its upcoming young elite who were to be sufficiently open to the transfer of new methods of management taken over from industry. 'Managed science' meant a goal-oriented approach in supporting a specific theoretical programme, that is, the reductionist programme within biochemistry and genetics; generous financial support for projects which included the improvement of technology as an integral part of and precondition for theoretical advances; support and encouragement for a collaborative research style and the expansion of a subtle, yet highly efficient, system of a patronage designed to intergrate young and promising talents into the new support programme. The deeper significance of the process of convergence between theoretical and strategic-institutional programmes can be grasped by analyzing the meta-language which was developed around it: in it certain key concepts were used both for the analysis of nature and its societal appropriation; with formal methods being used in analyzing living organisms as well as in organizing research.

The success of molecular biology by and large proves the efficiency of the organizational model that contributed to it.[10] The 'capitalization of life', as Yoxen has called it, became the new basis for biotechnology. Life itself is now seen as programmed and programmable and thereby undergoes a redefinition: it can be appropriated as intellectual property, it can be traded, bought and sold. Science as a commodity has taken on a new form and information becomes one of the keys for the organization of complexity.

The Discovery of Technological Sweetness: The Legacy of the Manhattan Project

The rationalization of the research process in accordance with new management principles and the converging organization of complexity, both on the theoretical-programmatic legal and in the comprehensive sense of organizing training and mobility requirements in a multinational chain spanning industry and universities alike, provide the wider setting of the forces that have decisively shaped the exterior conditions of what has been called the industrialization of science, or more recently, its 'collectivization'.[11] The internal changes, however, were no less marked: as the most far-reaching transformation they signal the change from the scientist working essentially as an individual to scientists working in a group. The new form of competitive cooperation which resulted from this seemingly innocent shift had a profound impact on the individual scientists' motivation, reward and career structures, self-image and sense of responsibility. The unifying process which directed this transformation was the shift from the production of discipline-based knowledge to the realization of science-based, technological projects – from science to science-based technological research.

98

Thus, what has sometimes been described as mission-oriented research is a much more profound change in the organizational mode of doing research than some have realized. Its history, as some as the methaphors still used convey, is war: its ancestor is the Manhattan Project.[12] There is not the space here to go into the detailed history of that project. Rather, I want to concentrate upon some of its structural legacies that appear to have come to stay. In the aftermath of its 'successful termination' – the dropping of the A-bombs – a more or less sincerely felt and profound moral discussion set in. What was now problematic were the altered conditions of the application of scientific knowledge; the loss of innocence of science; the changing relationship between the scientific and the political elites; but above all the moral implications for the individual scientist. What was not discussed, since it was not perceived as being problematic at the time,[13] were the side effects of the lasting organizational changes that turned out to be satisfactory and beneficient to both sides: to the scientific community and to the political-military establishment alike. The new mechanisms of scientific collaboration and masterminding science-based, technological research on an unprecedently by large and complex scale turned out to be too efficient not to remain. This efficiency has been refined and extended even though the particular historical conditions that gave rise to it have ceased to exist. Indeed, these mechanisms have been adapted to a smaller scale and they account to a large degree for the 'new breed' of scientists that we complain about as having lost openness.

Let us take a second look at what is known. One of the ingredients of the success of the Manhattan Project consisted in bringing together a group of highly qualified scientists and engineers from different backgrounds and disciplines – theoreticians, experimentalists, industrial engineers – and integrating them in a functioning, collaborative team. The towering influence of a strong personality was a decisive factor. Other ingredients of the successful model were: the strong moral and political appeal, which meant working under an urgent moral verdict to beat Nazi Germany in the race for superior weapons technology; the tremendous scientific interest aroused by the underlying theoretical problems; the fact that a segment of the scientific world elite was able to work together under unusual, immensely crowded but apparently satisfying conditions; the unquestioned disposal of almost limitless resources in highly concentrated form; and, perhaps most important, the direct concentration upon a specific task with a practical outcome: the bomb.[14]

Once this newly invented organizational model succeeded, admittedly under somewhat unusual conditions, it was too beautiful to let go. A number of institutions that were founded in the United States after the war incorporated the new organizational model, although there were heated discussions about the concrete form it should take: it was conceded that a certain amount of steering from the outside was both necessary and beneficial, but the well-kept illusion of autonomous scientists inside was also strongly maintained and the public funding of research became the determining factor of the new science policy. In its more mundane form, task orientation shed some of its military armour and blended well with the glamour of well-organized science and technology.

The superiority of the new organizational model lay precisely in its capacity to integrate different types of researchers, different skills and experience into some kind of optimal mix – always defined in view of a relatively short-term, concrete and practical task. The necessary correlation of increased individual mobility implied new career structures, in the sense of being able to move on to the 'better teams', ie the team with even more qualified researchers, more interesting problems and better-funded tasks. The inherent rewards were no longer defined simply by a reputation in one's own discipline, achieved, mainly, through discipline-based publications, but shifted towards the satisfaction of achieving a practical outcome, of being associated with a successful project. It became the company that lent its name to the product and being part of the company became an intrinsic reward.

But there were further repercussions of the Manhattan Project. Standing on the shoulders of giants and making one's own small contribution to the roster of claims towards immortality – which had been the aspiration of individual members of the scientific community for a long time – was also the hallmark of continuing a certain intellectual tradition. It was the hallmark of what I shall call 'noble science'. By contrast, the new scientific entrepreneurs, the condottieri of project-oriented science, were attracted by what one of their greatest representatives called the discovery of 'technological sweetness': the science-based, interdisciplinary orientation, set up and geared towards realizing a technological project that works. Its 'sweetness' conveys an attachment to the product as such. It is not seen in any larger context of application, but for its own, practical sense and which displays the quasi-erotic features of the automaton brought to functioning or life by the (male) scientist himself. It conveys something of the sensual pleasure that comes with the sense of total control, set up at the expense of narrowing the context in which the product is envisaged and made to work. Finally, sweetness contains also the invitation to try out, the offer of consumption – with a faint reminiscence of the paradisical apple.

Even if we know that the overwhelming majority of practising scientists and engineers worked and continue to work in conditions far removed from condottiere glamour, the appeal of technological sweetness is pervasive: we have but to take a glimpse at what Turkle in her studies of computer scientists has called the subjective side of the computer[15] in order to understand the intricate, compensatory mechanisms that constitute the technical appeal of their work routine which becomes transformed into something adventurous, non-routine and extraordinary when invested with the fantasies of omniscience.

Scientific Openness – A Disappearing Virtue

We need not reiterate here the fundamental contribution that scientific openness has made to the development of modern science; its inherent value in both a scientific and in a political-democratic sense. The example given by Ravetz – that it took a student to discover that the niobium used as an alloy in the steel of pressure vessels in a pressurized water reactor has such intense and long-lasting radiation that the decommissioning of such reactors will be enormously more difficult and expensive than previously assumed – cannot only be taken to ask what the scientific experts were doing who were supposed to check on such a possibility during the previous decades, but it is also a testimony to the still functioning openness in our institutional set-up, even in the realm of technology. Yet it may become a rare example, if the appropriation of intellectual property, its fragmentation and commercialization continue to be drawn into military secrecy by stages for the military domain already occupies almost 50 per cent of the world's total scientific and technological labour force.

It is worth taking yet another look backwards and recalling what some of the structural changes were that brought about the emergence and valuation of scientific openness. Initially, the norm of openness constituted a radical break with the secret tradition of the magus's trade. In Bacon's *New Atlantis*, the Mercatores Luci, who went out every twelve years in order to gather new scientific information of potential interest to the House of Salomon, went *incognito*. In another passage, we are told about the deliberations undertaken in order to decide which of the discoveries and knowledge of nature are to be published and which ones not. The members were bound by an oath of secrecy and although some things are revealed to the state, others are not.[16] We can assume that the economic context of incipient trade capitalism encouraged the free exchange of goods and the free exchange of scientific information alike, but the practice probably took longer to become established. It meant that the narrow and conventional boundaries of the craftmen's guilds and their inward-bound communication structure had to be transcended, and this occured only with the rise of the nation states. As long as the norms and behaviour of free exchange, in the widest economic sense, were thought to benefit the national interest, openness was encouraged and the universalist aspects of the message of the Enlightenment would receive enthusiastic support: the ideological and utilitarian components of openness would meet.

Today, with the new information technology at our disposal, scientific openness is challenged even on technolgical and social grounds. It is not merely that since science has become a commodity, so scientific information also has turned into a commodity to be bought and sold, guarded and destroyed. Access to it can be regulated and controlled in unprecedented ways and the impact of the new information technologies in doing so has been unduly neglected. It is not only secret information within the military research sector which has been separated and withdrawn from the ordinary network of scientific knowledge diffusion and circulation. The establishment of large-scale information systems with selective access implies a potentially deep structural change that may lead to a two-class system in access to scientific information.[17]

Faced with these and other developments which threaten scientific openness, what can we say about the responsibility of the individual scholar confronted with the decline of a disappearing virtue? First, we have to realize that virtues are not randomly distributed, nor are they something inherent or naturally given. Rather, they are tied to the emergence and decline of different social groups. Up to the Second World War science resembled in several of its structural features the former nobility: in its most visible activities at least, it was the preoccupation of a tightly inter-locking small elite, spread out across national boundaries and highly respected in its devotion to the pursuit of systematic knowledge and truth. Openness was a virtue linked to science as a noble activity. With the gradual and relative decline of science as such an activity, its corresponding virtues are also threatened and may disappear. For the new rising group engaged in it, the condottieri of project-oriented research, openness is dysfunctional. Competition between such groups gives advantages to those who know how to exclude others from information more often than by sharing it, at least at certain critical stages. Since publications – the former regulatory mechanism in establishing claims of priority and reputation and a strong structural support for openness – are also becoming obsolete or at least secondary in project-oriented research, legal regulations will eventually replace the self-regulation that has prevailed so far within the scientific community. In a curious reversal of Bacon's prophecy, the state will decide what is to be published and what not. In addition, with huge data banks becoming progressively more available, scientific information becomes – from the point of view of the individual scholar – externalized and reified in the sense that no individual can any longer keep up, let alone master the flow of information necessary for the pursuit of research. It is taken over, like many other activities in daily life, by a huge impersonal apparatus governed by the rules of access and retrieval, thus making it easier for the individual scholar to accept a contraction of openness.

But with science and technology occupying the place in the world they hold today, scientific openness has acquired a new and additional meaning. I refer to the changing relationship between science, technology and the public. With the actual decline of scientific openness as an internal regulatory mechanism, the ascendency of openness as an external relationship directed towards the public may even gain in additional importance and urgency. In the last decades we have witnessed controversies around large-scale technological developments and we are about to undergo another deep crisis about nuclear weapons which undoubtedly will affect the public's attitude towards science and technology in general. Openness in this field of conflict does not mean to engage in the kind of pseudo-public relations activities that have served as an inadequate surrogate in the last few years. Rather it is a serious challenge, calling for an institutionalized pattern of honest and open exchange of information on matters of legitimate public concern. I think it is in this area where there is still much room for individual responsibility and collective responsibility alike. Furthermore, openness in this second sense is not only compatible with the new style of scientific research, but becomes mandatory if this style is to survive a growing distrust of the social and political consequences of the results of this kind of research. It is therefore also in the self-interest of science as project-oriented research to expand its responsibilty in the direction of working towards a more trustworthy public basis.

Scientific openness, even if it may be a disappearing virtue that will fade eventually together with the world of noble science, may remain with us in a broader, more democratic and urgent sense.

Can Good Men Fail?

We have come full circle to our present predicament which is that we know that in science and technology too, things can go wrong. The old tenet of noble science was that individuals can fail, but that their errors, their frailties, even their personal weakness, will be compensated by the collective march of scientific progress. The self-correcting mechanisms within science as an institution were deemed to be sufficiently strong and efficient to guarantee that despite minor aberration the deviation on the route to progress would never become sufficiently large to allow a serious derailment. However, this faith in a collective venture which has rooted in the working experience of academic scientists, can no longer be maintained in face of fragmentation into highly mobile, decentralized and transient research units, which are constituted and dissolved in accordance with the life cycle of a multitude of projects. We have come to realize that it is neither sufficient to presume that good men alone will do good science – in all the various meanings of the word – nor is science alone capable of producing good men whether inside or outside science. And anyway, where are the women?

Would science be any different, if women were included in sufficient numbers, would it have a more human face? Like any hypothetical case it is easy to argue and impossible to prove. I must admit, however, that it is hard to imagine that women would be as easily seduced as men by technological sweetness; that they could as readily be drawn into the 'alchemy of the arms race' as S. Zuckerman has called it; that their blindness *vis-à-vis* potential global disaster would be identical to the blindness of their male colleagues, rather than be of a different kind. But it can hardly be expected that the presence of women in positions now occupied by men would of itself change the structure of science. The rules of the present game would have to be changed radically. But even then, good women – even if they have other, not yet tapped human resources than men – are also human; they too can fail.

Where does this leave us finally? If we ask what has happened, the answers are easy to give: technological sweetness has turned sour; war, although a sad constant in the conduct of human affairs, has become an apocalyptic vision with the help of science and technology and what Ravetz calls technological blunders are becoming increasingly menacing. It is much more difficult to answer the question why. There are no obvious scapegoats to blame, no single agency or institution, let alone groups of individuals. Rather, we are confronted with a large-scale historical process, a blind historical process, directed, but not planned as such, in which we all participate. In its unfolding, science has gained the means of changing, even annihilating the world. As other monopoly holders have experienced in other fields, the successful acquisition of the monopoly has changed the nature of science.

It is the transition from science as a noble, academic activity, which was mainly university-based, to a science-based technology and a project-oriented research-science that has been the focus of my exposition. In this process of transformation, scientific openness in the traditional sense is likely to fade out, while being replaced by scientific openness as a new and yet to be institutionalized challenge. Good science and good men are insufficient. Good women have had no chance, as yet, to alter the rules of the game. Nor is it sufficient to believe any more that only the social and political order has to be changed if good science is to have a chance. The convergence of scientific activity with profit-orientation, the bureaucratization and the militarization of science have been the topic of this book. But an appeal – even a strong moral appeal – to individual responsibility alone will not do either.

If good men can fail – which is a very likely possibility – what we need are better organizations. This means deliberate designs, constructed with a view to the traits we value and wish to preserve. It is a call for our collective imagination to envisage what such organizational designs could look like, based on a thorough understanding of the functioning of organized scientific life operating within a social and political context. Nor is this a task unique to science. After all, it has taken centuries and many struggles until political institutions have been devised that allow for the play of checks and balances that constitute the central tenets of our democratic institutions.

One important question still remains, however: how can organizations know what is good science? Or would we opt for less?

Acknowledgements

My sincere thanks go to Wolfgang Reiter who spent a long, sunny Sunday afternoon with me in a Viennese coffee-house, discussing the Manhattan Project and its aftermath.

References

1 Sprat Thomas 1667: The History of the Royal Society of London.
2 Ravetz J. 1982 The Social Functions of Science: a commemoration of J. D. Bernal's vision. Science and Public Policy October, pp 262–266.
3 Neurath Otto 1979: Wissenschaftliche Weltauffassung, Sozialismus und Logischer Empirismus, Rainer Hegselmann (ed.). Suhrkamp, Frankfurt.
4 Mendelsohn Everett, Nowotny Helga (eds.) 1984 Science and Utopia. Yearbook in the Sociology of the Sciences. Reidel, Dordrecht, vol 8.
5 Oppenheimer Robert, quoted in Nuel Pharr Davies, Lawrence and Oppenheimer, New York, 1968, p 16.
6 Ravetz J. 1982: op cit.
7 Latour Bruno, Wollgar Steve 1979: Laboratory Life, The Social Construction of Scientific Facts. Sage Publications, Beverly Hills.
8 See my critique of these studies and further ideas on which this section is based. Nowotny Helga 1982 Leben im Labor und Draußen: Wissenschaft ohne Wissen? Soziale Welt 33 (2) pp 208–220.
9 See especially Kohler Robert 1976 The Management of Science: The Experience of Warren Weaver and the Rockefeller Foundation Programme in Molecular Biology. Minerva 14: pp 279–306 and Yoxen Edward 1981 Life as a Productive Force: Capitalising the Science and Technology of Molecular Biology. In Levidow Les, Young Bob: Science, Technology and the Labour Process. CSE Books, pp 66–122.
10 A recent interesting reassessment of the role played by Warren Weaver's management philosophy corrects details, but does not, in my opinion, contradict the thesis advanced here. See Abir-am Pnina 1982 The Discourse of Physical Power and Biological Knowledge in the 1930s: A Reappraisal of the Rockefeller Foundation's Policy in Molecular Biology. Social Studies of Science, 12 (3) August: pp 341–382.
11 Ziman John 1983 The Collectivisation of Science. Bernal Lecture to the Royal Society, April (mimeo); which came to my notice only after this paper was written.
12 There exists, of course, a vast literature on this subject. Among others, see Kevles Daniel 1971: The Physicists. Vintage Book, New York; Grabner Ingo, Reiter Wolfgang 1982 Ende und Fortschritt der Physik. Vienna, (mimeo).
13 This is not to say that research policies were not discussed; quite the contrary. However, the main aim was how to advance the successful organization model and not to reflect upon its internal consequences. See Kevles Daniel: op cit.
14 Some have referred to this as a highly pressurized monastic life. According to Daniel Kevles, 'Los Alamos scientists skied and hiked in the remote beauty of the surrounding mountains, gathered at intellectually stimulating social evenings and kept the maternity wards busy'. Op. cit., p 330.
15 Turkle Sherry 1982 The Subjective Computer. Social Studies of Science, 12 (2) May: pp 173–206.

16 Bacon Francis and New Atlantis. In The Essays Odham Press, London, p 411.
17 A case in point, is the Advanced Research Project Agency, access to which is open to certain institutions in NATO countries. On the whole, the 'information war' going on behind the closed screens of computer terminals has been neglected by social studies of science.

A New Branch of Science, Inc.

The Separation of Science from Public Policy

Those responsible for scientific policy occasionally run the risk that a piece of unanticipated reality may be lurking behind the metaphorical imagery they have constructed in order to accommodate a broad spectrum of different ideas. As the organizers of the Science for Public Policy Forum remarked, the conventional link between science and public policy is to think in terms of public policy for science – a long-standing concern among a small circle of experts drawn from the natural sciences, the policy sciences, and politicians as to how to find optimal ways of funding research and of guiding the innovative process of scientific-technological development. The Forum organizers asked us, however, to consider the converse combination: science for public policy. This has, as I will try to show, both an obvious ring of familiarity, asking us to restate and perhaps clarify the directive mission contained in the pronoun, but at the same time a more provocative meaning inviting us to overcome the *de facto* separation of science from public policy.

Let me first consider the obvious meaning: science for public policy as the outgrowth of the oldest social mission of science – for the public good. Ever since the inception of modern science in seventeenth-century England, with the incisive formulations of Francis Bacon, scientists and technologists have conceived their activities in terms of noble aspirations. By linking their work to an increase in welfare – first of their own nations, but ultimately of the entire human race – they sought to reduce suffering due to the lack of means, to satisfy material wants, and to alleviate degrading labour. The collective purpose of science conceived in these broad terms has hardly changed. In the latter part of the twentieth century the common good is still on the public agenda and policies are still directed towards tangible results. As Harvey Brooks has reminded us, the standard list of fundamental human needs to which science and technology are expected to contribute is still remarkably unchanged: food and energy supply, health needs, transportation, shelter, personal security. Later additions seem to be the remaining items: a cleaner environment and a social system which, in the words of Harvey Brooks, facilitates rapid adaptive change while restraining the possibility of violent conflict.[1]

Reprinted with permission from Brooks H. and Copper C. L. (eds). Science for Public Policy, Pergamon Press 1987.

Such additions to the standard list already signal the shift from the tangible results of science and technology, from their expected direct contributions to economic growth and welfare, to the more intangible, indirect, and mediated ones. Today, science for public policy can no longer concentrate on accelerating the rate of innovation as an aim in itself. Rather, it has increasingly become preoccupied in dealing with the unwanted and unintended effects of its direct contributions. The quest for a cleaner, safer environment is a case in point. The secondary and tertiary effects, the as-yet unknown consequences, of our interaction with the environment have become the source of our main concerns. There is an equal quest for a social system that would facilitate adaptive change and yet not be overturned by it. The expected contributions of science for public policy have shifted from the operational to the symbolic realm. Utilizing its cognitive capacities, putting knowledge as the most precious resource science has to offer at the disposal of policy-makers thought to be in desperate need of it, scientific knowledge and information has become the key for managing a future whose existence is threatened by the interventions made in the past. Science, so far, holds an absolute monopoly on this kind of knowledge and, as other previous monopoly holders, it had to maintain its claims by guarding its institutional boundaries, in this case, its autonomy in the production of knowledge. This is one reason why the dividing line, separating scientific facts from values, ordinary everyday knowledge from scientific knowledge, scientific expertise from lay participation, and science from politics, is so entrenched. What science had to offer – according to its own definition of its social mission – was advice: advice held to be clean from political considerations, free from values and mere opinions, from interests and control over its later applications. Science was disinterested and neutral, committed solely to its own impartial and context-independent conception of Truth. This, at least, was the ideal.

But is such a formula sufficient? Is this what science for public policy is all about, when the pressure of taking action mounts in areas of genuine scientific uncertainty, and when the roles of what once were thought to be "hard" scientific facts amid "soft" human decision-making procedures, as Jerome Ravetz has pointed out[2], are becoming reversed and we now are confronted with the necessity for making "hard" decisions in the face of "soft" scientific evidence? While science for public policy is firmly ingrained in the social mission of science, both in the sense of tangible, instrumental results and the more intangible resource of providing information for guiding the policy processes, the lines separating science *from* public policy are also sharply drawn. Harvey Brooks states this very clearly: science and technology, he writes, cannot provide a solution by themselves. They can only generate the conditions in which a society can develop a solution.[3]

But does the policy process really live up to the expectations put into it? Who does the translation from one field to the other in the first case and what happens (as invariably it does happen) if scientific findings get transformed, distorted, subject to political bargaining in the translation process? Is it really true that science 'only' creates the conditions in which society can develop a solution? Are not both science and the evolution of an institutional societal framework geared towards the production of certain types of solutions, linked to each other through a common historical ancestry? Are not both, as Max Weber suggested a long time ago, embedded in the process of ongoing rationalization that happened to be both a precondition and the most important consequence for capitalism to evolve, bent on achieving a high degree of predictability and calculability, of efficiency in the domain of nature as well as within the social and economic order? While the spillover effect of the scientification of everyday life, including political institutions, has been enormous, one ought not to lose sight of the tremendous changes that science, its organization, and the concept of science have undergone in this very same process.

Thus, the innocuous looking line that restates the obvious – that science is for public policy – while at the same time separating science from public policy – by claiming that it only creates the conditions for society to develop solutions – opens up a dilemma which is becoming more acute under the pressure for new solutions on the part of science for public policy.

How Rational is the Policy Process?

The impact of the process of rationalization has been uneven: while the organization of scientific knowledge became the model of rational organization *per se,* the political process is generally viewed as lagging far behind. It is worthwhile to recall the great appeal that the scientific method once commanded as a way of settling disputes, and the futile hope that was expressed again and again, in scientific and political utopias alike, that it would be possible to arrive at similar rational procedures for solving conflicts in the political realm.[4] The dominant view of science for public policy shares some of these elements, since it rests on the implicit assumption of an underlying structural similarity of mutually converging rationalities. This assumption has been elaborated in two directions: one is the still dominant model of rational decision-making that was devised especially by policy analysts, and the other one is the view that a great number of scientists hold about the nature of their input into the policy process. This picture of a rational or, perhaps better, over-rationalizing model of the policy process has not failed to repeatedly attract well-founded criticism. Majone, among others, has pointed to an underlying deeper commitment to a teleological, end-result conception of policy-making and the reliance upon a number of fictional constructs which follow from the model.[5] In a thorough review, Aaron Wildavsky highlights the essential difference that exists between puzzles – to be solved once and for all – and (policy) problems that may be alleviated, eventually superseded, and finally redefined. He declares that the "rational paradigm" is simply mistaken. It fails to adapt to the ways in which decisions are actually made, where available answers determine the kinds of questions that are asked and objectives are never the products of the seat of rationality, but dependent upon available resources.[6] Others, like Peter House, have systematically questioned the assumptions by which policy analysis was supposed to be brought into the policy process, by comparing a number of actual cases with their analytical foundations.[7] In attempting to explain why policy-oriented research seems to have had little or no direct impact on policy-making, Björn Wittrock has suggested that the mismatch between the supply and the use of policy-relevant social knowledge can be traced either to a highly rationalistic conception of the policy process – the "social engineering" model – or to an "enlightenment" model that assumes that social science research does not so much solve problems as provide an intellectual setting of concepts, orientations, and empirical generalizations. He argues in favour of a third model – a dispositional one – a conception of knowledge utilization: the process is neither arbitrary and haphazard, nor entirely pre-programmed; important policy research must be there to be utilized and if conditions are propitious and important actors available, its findings might well have an impact.[8]

While some of these commentaries and criticisms pertain more to the utilization of social science knowledge, there is widespread recognition of the enduring and conflicting nature of public policy issues in general which have increasingly come to include environmental and technological issues.[9] In such an enlarged view of policy analysis, the question of the epistemological foundation is also receiving renewed attention. Thus, in a recent review of policy research and a rejoinder undertaken in defence of the policy sciences as science, one consistent theme of contention between the authors was that one of their models would follow an outdated positivistic conception of science, while the policy-making process should be viewed as resting on a much broader epistemological basis.[10]

While some of this ongoing dismantling of the Received View can be interpreted as a necessary correction of the immature field of policy analysis, I think that the reasons lie somewhat deeper. The Received View has been adopted not only by its proponents – over-confident about rational problem solving and about the extension of methods and tools from one realm – that of military and industrial operations – to the much more complex and ambiguous arena of political and social issues – but also has its adherents among actual decision-makers and scientists alike. It conformed to the Enlightened View that science and public policy were either slowly converging in their inherent rationalities or that public policy, in order to be receptive to scientific advice and improvement, had to come to resemble more closely what a scientific model of the policy process demanded it to be. This was a highly convenient way of thinking about science for public policy, as long as it remained the exclusive concern of a relatively small circle of public policy officials and scientists involved as advisors in certain policy arenas. It fitted into an institutional arrangement, moreover, that defined public policy as falling within the competence of a relatively closed administrative-scientific coalition.

Not surprisingly, the correlative view held by many scientists involved in the policy-making process as experts or advisors carries an equally strong faith as to what good public policy is all about. It is to be guided by scientific expertise which claims authority also over the definition of good government: one that admits to strong scientific guidance in how to conduct political affairs. There was a recent reanalysis of the testimony of some 130 expert witnesses who stated their views on the necessity and desirability of creating a US Congressional Office for Technology Assessment. Most of these witnesses were of the opinion that technology is to be equated with effective intelligence which they considered to function as a substitute for an otherwise failed sense of history, of logic and purpose in the unfolding of events.[11]

Although expressed in a particular context and referring explicitly only to technological expertise, such views probably accurately reflect the confident attitude of a scientific-technological élite involved in the public policy process so long as their equally held belief in the impartiality of their expertise remained unchallenged. Good science and good public policy would meet as long as both would conform to the underlying assumption of the growing convergence in their rationality. The shock and disturbance which came with public contestation were accordingly great.

Science Contested: Science for Whom?

As long as public trust in science and technology was still high and undisturbed, as long as it was a small circle of a scientific élite that functioned as advisors to governments and administrative officials, as long as the public image of science would reasonably cover what scientists themselves projected their activities to be for society – science for public policy was what good scientists did for a rational policy process.[12] The internal hierarchy of the status system was sufficiently strong to carry its weight in the public arena and the internal status system determined who a good scientist was. Looking back to the time before public contestation seems almost like looking back at a bygone age. Science and public policy have long since ceased to be bound by a relationship consisting simply of a few representatives of science and a few policy-makers and officials. The public has intruded in public policy and is, so it seems, here to stay, even though it is not always easy to say who the public is. Most observers would most probably agree that a new set of political actors and new social movements have come to the fore in the stream of an altered public awareness of the impact of science and technology. They have done so, first by questioning what has been taken for granted so far, namely that science always works for better public policy; then by protesting that their concerns were not taken into account properly; and finally by claiming that science for public policy should be subject to participatory scrutiny like other inputs into the political process. Since it had become obvious that science and technology could sometimes have negative side effects and even potentially cause great harm, the assumption valid from the seventeenth century onwards that science would inevitably produce results for the public good has definitely come to its end.

Among the many repercussions that the protest phase had on the relationship between science and public policy, I want to single out those that illustrate the changes of the context in which science for public policy is defined today. This changed context reflects a new balance of the tensions inherent between science and public policy.

The first outcome is the undermining of the alleged rationality of the political process, which turned out to be far less rational than depicted by the champions of rational policy analysis. There was not one unitary decision-maker but a multitude of conflicting parties. The political process showed itself to a certain degree receptive to protests, and new forms of political intervention were designed to distort, disrupt, and alter the way politics was routinely conducted. The high standards of rational decision-making quickly dissolved under the eyes of the empirical observer, yielding their place to a mixture of power games, arduous negotiation processes of political bargaining, and recourse to already institutionalized conflict-solving mechanisms, such as the courts. While nothing in this is surprising to political scientists, it came as a surprise nevertheless to those who had thought that scientific advice was exempt from these ordinary forms of political rationality. When confronted with scientific advice and expertise, the policy process did not display the rationality expected. Scientific expertise was treated like any other input into the political process: as a political resource to be used by both sides, negotiable, and not necessarily 'true'; in any case not endowed with higher political credibility than other inputs.

The second outcome is related to the first. It underlines the inherent difficulty in reconciling the idea of scientific knowledge, generated in accordance with its own methodological canon of objectivity and intersubjective validation, with demands of popular participation. What can be shared to a certain extent – 'popularized' as the term has it – comes after scientific facts have been established and a body of knowledge validated. It is the diffusion of knowledge and, to some extent perhaps, its application that can be opened to public participation, but not the process of scientific knowledge as such. Yet, in the public contestation phase, the objective findings of facts, their precondition as well as political consequences, were challenged. Thriving on the open disagreement of experts in public, a more transparent model of science for public policy was proposed, an adversary system that would allow for some kind of representational system of comparing scientific findings and methods of arriving at them. By juxtaposing experts and counter-experts, each chosen as trustworthy from the opposing parties, science was to become more democratic. Underlying such a proposal was of course the expression of a deeply-seated distrust of science functioning as an objective enterprise and standing above vested interests. In the public contestation, science was charged with taking sides with other powerful interest groups in society and therefore discredited as not being truly for public policy. The other two changes affecting the dominant conception of science for public policy arose out of internal reflection and critical evaluation, notably through sociological studies of science. They show science not to be as neutral, objective, and free of social interests as the positivistic ideal of science affirmed for a long time, and claim that all scientific knowledge is socially constructed and negotiated.[13] Scientists were shown (in their own accounts of how they arrived at results) to oscillate between a usually informal context of contingency, in which they admit the uncertainty and provisional nature of the knowledge in question, and an empirical, formal context in which they justify the conclusions reached by emphasizing solely the certainty and absoluteness of the results they obtained.[14] Both of these themes represent revisions of the official model of science, the standard model confirmed by the public rhetoric of science. Although the critical dismantling of some of its features came from inside science, so did public controversies throw open the not-so-objective sides of objectivity and add the weight of context-dependency to the process of scientific inquiry. Among others, Brian Wynne has noted that it is important to see clearly that such criticisms and invitations to self-reflection are not to be taken as an all-out assault on science; nor is it a question of deliberate bias and wilful distortion on the part of scientists that needs to be publicly exposed. Rather, the all-pervasive message of such studies and detailed critiques is to make a much more general point: that the definition of a scientific problem is never isolated from the political context in which it occurs, nor can political implications be completely eliminated from the course of the analysis and policy conclusions derived only at the end.[15] Put in another way, I would add: we have to recognize and accept that all scientific analyses tied to a given policy context anticipates and reacts to the often unstated assumptions of policy outcomes. The use of concepts, the substantive implications of methodological procedures, the utilization of any kind of data cannot but be impregnated with different policy meanings. To claim anything else would be utterly naive and could not be upheld in the face of overwhelming empirical evidence to the contrary. How to utilize this knowledge for better public policy purposes is, however, still another matter.

In the period of public contestation and its aftermath, science for policy has been turned into the question of 'science for whom?'. While the policy arena has been potentially enlarged by a wider public that wanted to be heard, the lessons to be drawn from the demystification of the over-rationalized political process and the over-rationalized image of the internal workings of science are by no means clear. If we admit that policy-prone types of scientific analysis inevitably bear the marks of their contexts of justification, of contingency, and of political relevance; if we admit that the informal process of scientific reasoning, of the utilization of data, and their interpretation include much stronger doses of intuitive judgement, implicit values, and tacit procedures of persuasion – are we set on a course which leads not directly to hell, but to something akin, namely scientific relativism? Or, as many scientists (who still uphold the ideal of no science-in-public) would maintain, would a greater degree of honesty and modesty about the internal workings of the scientific process lead only to a further decline in public trust or increase public apprehensions, perhaps wilfully distorted even further by the media? Is there a way out from haughty retreat behind a formal position and from apologetic relativism alike?

Between Orthodoxy and Reformism

The orthodox response has been to reassert the traditional separation of science from public policy, arguing that only then can it be *science* for public policy. Similar statements abound in the policy field dealing with risk analysis, risk assessment, and risk management. A recent study prepared by the National Research Council of the US Academy of Sciences makes an explicit distinction between risk assessment and risk management: risk assessment is to be based on scientific judgement alone and has to find out what the problems are; it should therefore be protected from political influence. Risk management, on the other hand, is defined as the process of deciding what to do about the problems. It involves a much broader array of disciplines and is aimed towards a decision about control.[16] Perhaps more clearly than other policy studies, risk analysis has been confronted with the problematic situation that is inherently at the heart of most of them: while the intention is to provide as clear and careful a basis for action as possible by diligent scientific scrutiny of the hazards that can be subject to analysis, the selection and implementation of intervention measures generally involve balancing scarce resources, political goals, changing social values, and sometimes a somewhat unpredictable public opinion.[17] Another study published by a group of the UK Royal Society, equally devoted to methods and approaches to risk analysis, reached a different conclusion in which the whole process, including risk estimation, risk evaluation, judgements on acceptability, and taking account of public opinion, is referred to as risk management.[18] The respective roles of these two parts of the process are treated differently.

The chances for a successful application of the relativistic strategy are even slimmer. Not only is relativism a highly contested philosophical position within the theory of science,[19] it has few, if any, friends among practising scientists. Even if we would leave aside the deeper philosophical issues and concentrate on a reformist plea for greater public openness about the internal side of science in which subjective judgements have their place, uncertainty abounds, and room is even made for errors – would this alone provide a better basis in the face of pressure for political action when confronted with incomplete and uncertain scientific knowledge? Although the public image of science is in urgent need of correction in the reformist vein, no miracles can be expected from this strategy if nothing else changes.

This takes us back to the questions raised at the outset of this chapter. If science only creates the conditions in which a society develops solutions, we may ask from a sociological point of view which kind of solutions are likely to emerge. If science itself takes proper notice of the increasingly recognized realm of uncertainty, due not only to the human condition of ignorance but to the knowledge gained about the interacting secondary and tertiary effects of scientific and technological interventions in the natural and social environments, the conditions are created for science *and* society to develop new kinds of solutions. On the epistemological side, this can be an intellectually exciting venture; for the policy process it might reveal some unexpected results.

So far, the historical conditions have favoured one particular type of solution: the utilitarian-instrumental one. Utilitarian solutions have pressed for the increased applicability of scientific knowledge, for its industrialization and more efficient organizational forms, and for its relevance to continued innovation. The concomitant societal mechanism aiming for the distribution of the surplus thus created, for motivation of the work force, and for the smooth functioning of societal institutions has been an instrumental type of rationality concerned only with efficient and hierarchical means-ends procedures that have become the guiding principle of how social affairs are conducted in the industrialized West. Yet, we have also come to realize recently that the conditions created by science and technology have increasingly cast doubt on the adequacy of these solutions as a guide for policy. The discussions about accelerated economic growth in the face of environmental damage and the threat to the overall balance between nature and man have been only one facet of growing uneasiness. Discussions within the scientific community on how to cope with uncertainty under the outside pressure for action have underlined the limits of the utilitarian-instrumental solution.

The Rise of the Managerial Conception of Science for Public Policy

The utilitarian-instrumental solution allowed for a clear-cut separation of science from policy while maintaining at the same time a strong (utilitarian) link of science for public policy, based on a means-ends relationship. While the production of scientific knowledge needed its autonomous space, it was assumed that it would lead more or less automatically to its social utilization since this was the in-built direction for scientific technological development to take. Steering clear of too-close a contact with the political system, "not meddling in politics", science became closely enmeshed with the industrialization process and its aftermath.

Science is now confronted with new demands from the political process. As with the industrial system, the question is not so much one of direct influence or control. The scientific system has guarded surprisingly well the core of its institutional autonomy. It was at the height of industrialization in the latter part of the nineteenth century that major industries in Europe became science-based, and the split between basic and applied science was successively introduced. I see something similar occurring today, with science yielding to the powerful and all-pervasive political context that demands new scientific solutions for dealing with problems that science and technology have helped to create. An institutional split – which is also epistemological, concerning methodologies, substantive content, and professional self-understanding alike – is likely to occur within the sciences – between a public policy branch and an academic branch. But there is no ready-made kit of tools and recipes, of techniques, nor computer simulation models which can easily be drawn upon to fill the knowledge gap. Rather, the epistemological and practical basis for this latest branch in the differentiation of the sciences is yet to be created. In order to be successful, it has to have a strong epistemological tradition within at least some of the sciences themselves; it has to be ideologically attractive; it has to be politically feasible. It has to hold out the promise of conceptual power and clarity and, at least, a methodological armory that is adequate for the types of problem to be addressed. In

116

short, it has to embody a vision of being able to meet the demands of the policy process without relinquishing its own social and cognitive identity, and without giving up its strong claims to institutional autonomy from direct political interference. In order to keep its position as monopoly holder of the most cherished type of knowledge and to be trusted by the public, confidence in its impartiality has to be restored. These criteria are met by a new conception of science for public policy which I call the managerial conception of science.

The development of the managerial conception occurred gradually and on several levels. At the height of environmental concerns, when the limits of growth and exploitation of natural resources became a newly perceived part of reality, resources were suddenly seen to be finite – to be managed for the interest of all. When technologies were threatening to get out of hand and in urgent need of new kinds of control, we started to speak of managing them. When it became clear that the new problems created through scientific-technological interventions, with their unknown, unintended, yet potentially harmful effects, could not be solved in the accustomed way – if ever at all – we switched in our rhetoric from solving problems to managing them. This is a reasonable adaptation to a new situation in which too many variables were interacting under highly uncertain temporal conditions and in which the resilience or robustness of systems had yet to be determined empirically and theoretically. The thought of management comes easily to systems thinking, as this is one of its more precisely defined roots.

The managerial conception of science for policy also contains an implicit plea for shared responsibility at a time when individual responsibility has lost all ground in the modern organization of science. It is no coincidence that it alludes to a corporate style: management of problems which cannot be solved; management of uncertainty rather than a quick and unfounded (irresponsible) hope that it disappear quickly. This contains an appeal to a multi-leveled hierarchy of responsibility adequate for the new kind of situation we are facing. In contrast to a notion like "muddling through" which Charles Lindblom proposed, with very moderate success, to explain the political process, the scientific management of problems proclaims a relatively high degree of control in the face of a sea of external uncertainties. It contains the promise of exploiting new opportunities, should they arise, and of ways to "identify and carry out actions that will allow us to change the rules of the game".[20] In short, management, and especially scientific management, is a respectable, orderly procedure with a high degree of success in economic life, particularly within large-scale organizations. It implies a certain type of rational behaviour since it is goal-oriented, but also takes account of unavoidable constraints. It has a formal and an informal side, as every student of organizational behaviour knows and good management is apt to utilize both to the fullest. Contrary to the political model of accountability, defined as the electorate in Western democracies, managerial accountability rests on the assumption of a built-in hierarchical structure of duties and liabilities which is only ultimately responsible to a distant and abstract entity (the 'owners') who are not supposed to interfere. Thus, one of the strong appeals of the managerial model over a kind of political model lies in the high degree of autonomy it promises to the managers – in this case, to scientists. While it has remained problematic to defend the autonomy of science in the face of its role in the political process, the managerial conception promises a way out: while not denying the need for a built-in system of responsibility, its exact nature remains shrouded in a veil of competence in the double sense of the word; competence of those who are capable to handle scientific policy matters and of those who are officially charged with handling them.

The new conception of science for public policy – as distinct from academic science research – reduces the old question of how to maintain the boundary between science and public policy to irrelevancy, since by definition scientific management of policy problems stands above the need to protect science from political intrusion. It has all the evocative power of a new mediating institution and of a new social invention in the face of otherwise unsurpassable contradictions. It is an elegant solution and I predict that it will work successfully. It can incorporate the orthodox response and the reformist strategy described above: the former by interpreting the protective line being drawn between scientific fact-finding and political decision-making as being merely an administrative procedure; the latter by proclaiming greater honesty about inherent biases in the way science works as being part of the informal side of the management process.

The new ethos of science for public policy will be that of scientific managers, and good management is for the sake of the company. The only drawback I see is the question that remains open: who is the company and who controls it?

References

1   Brooks, H. (1981): Some notes on the fear and distrust of science, in A. S. Markovits and K. W. Deutsch, (Eds), Fear of Science – Trust in Science (Cambridge, USA: Oelschläger. Gunn. and Hain Publishers).
2   Ravetz, J. R. (1985): Uncertainty, ignorance and policy, this volume, Chapter 7.
3   Brooks, H. op. cit.
4   Mendelsohn, E. and Nowotny, H. (Eds) (1984): Science between Utopia and Dystopia: Yearbook in the Sociology of the Sciences, Vol. 8 (Dordrecht, The Netherlands: Reidel).
5   Mojone, G. (1981): Shortcomings of the Policy Science Approach to the Analysis of the Public Sector, in F. X. Kaufmann, G. Mojone and V. Ostrom (Eds.), Guidance, Control and Evaluation in the Public Sector (Berlin, FRG: Walter de Gruyter).
6   Wildavsky, A. (1979): Speaking Truth to Power (Boston, USA: Little, Brown, and Co.).
7   House, P. (1982): The Art of Public Policy Analysis (Beverly Hills, USA: Sage).
8   Wittrock, B. (1983): Policy Analysis and Policy-Making: Towards a Dispositional Model of the University/Government Interface, Report No. 29 (Stockholm, Sweden: University of Stockholm, Sweden, Group for the Study of Higher Education and Research Policy).

9 Coates, J. (1978): What is a public policy issue? In: Judgements and Decision in Public Policy Formulation (Washington, DC, USA: American Association for the Advancement of Science Selected Symposium 1) pp. 34–69.

10 Schneider, J., Stevens, N., and Tornatzky, L. (1982): Policy research and analysis: an empirical profile, 1975–1980, Policy Sciences, 15, 99–114; Brunner, R. (1982) The policy sciences as science, Policy Sciences, 15, 115–135. See also Brewer, G. and de Leon, P. (1983) The Foundations of Policy Analysis (Homewood, USA: Dorsey).

11 Doughty Fries, S. (1983): Expertise against politics: technology as ideology on Capitol Hill, 1966–1972, Science, Technology, and Human Values (Spring).

12 Nowotny, H. (1984): Does it only need good men to do good science?, in: Science as Commodity, M. Gibbons and B. Wittrock (Eds) (London, UK: Longman).

13 A good sampling of the literature can be obtained, in: Social Studies of Science.

14 Mulkay, M. (1983): Scientists theory talk, The Canadian Journal of Sociology 8 (2, Spring).

15 Wyne, B. (1983): Models, Muddles and Megapolicies: the IIASA Energy Study as an Example of Science for Public Policy, Working Paper WP–83–127 Laxenburg, Austria: International Institute for Applied Systems Analysis.

16 National Research Council (1983): Risk Assessment in the Federal Government: Managing the Process (Washington DC, USA: National Academy Press); Ruckelshaus, W. P. (1983): Science, risk and public policy, Science 221, 1026–1028.

17 Coppock, R. (1983): The Integration of Physio-technical and Socio-physical Elements in the Management of Technological Hazards, Mimeo (Berlin, FRG: Science Center, International Institute for Environment and Society).

18 The Royal Society (1983): Risk Assessment. A Study Group Report (London, UK: The Royal Society).

19 For a glimpse of an ongoing debate see Roll-Hansen, N. (1983): The death of spontaneous generation and the birth of the gene: Two case studies of relativism, Social Studies of Science, 13, 481–519.

20 Clark, W. C. (1980): Witches, Floods, and Wonder Drugs – Historical Perspectives on Risk Management, R–22 (University of British Columbia, Canada: Institute of Resource Ecology).

The Difficult Dialogue:
The Social Sciences in Search of
Useable Knowledge

Social Science Research in a Changing Policy Context

Social Science Research – A Commodity?

To speak about producers and users of social science research means treading on economic territory, at least in the metaphorical domain. It is a language that conjures up images of demand and supply; that deals with activities like buying and selling. Production and use, i. e. consumption also implies the image of a market on which certain kinds of goods can be sold easily, while others may have to sell below their real value. Some goods rapidly become obsolete and for others there seems to be no marketable outlet at all. It implies a contractual relationship with the pressure of deadlines and the prospect of either renewal or termination. In short, we treat research as though it were a commodity . . .

Social scientists, even the most hard-nosed and technocratically oriented ones (and even those who work in commercialized research units) have never been entirely comfortable with the image just evoked. For one of the strongest underlying assumptions of social science research has always been that of intended usefulness. The often covert ethical tenet is one of public service, one that dates back to the old wish of serving as advisor to the Prince.[1] The assumption was not entirely unfounded. There was a time – not too long ago – when many of us were actually lured by the signs that the princes of the day sent out – signals for wanting our advice: in the planning euphoria of the late 1960s and early 1970s the belief in the malleability of societies and in rational planning as an indispensable feature of mature societies not only permeated our text-books, but also led to an unprecedented flourishing of applied social science research. It was carried by the belief that societies could be reformed through the fiat of political interventions and our belief in the possibility that we as researchers could produce solutions to any kind of problem posed to us.

But, as Brecht already remarked, belief is often followed by doubt. In the worst case, it is followed by resignation, by academic retreat and by a new kind of defeatism. But the period of sober re-awakening that followed the period of euphoric expansion, has also offered us a chance to re-analyze what went wrong. Several explanations have come to the fore. They range from simple 'translation errors',

Reprinted with permission from Nowotny H, and Lambiri–Dimaki G. (eds.), The Diffucult Dialogue between Producers and Users of Social Science Research, European Centre for Social Training and Research, Vienna 1985.

i. e. a deficiency in the communication process which may be faulted either by not adequately recognizing its proper structure and its imperatives, or by not being able, on the part of the researcher, to ply its exigencies. The time dimension of research projects does not easily dovetail with the temporal demands of policy-makers, and policy cycles are often out of tune with regard to circles of academic fads and fashion. Other well-known traps contained in the communication process are the language that is being used and the political contingencies that remain uncontrollable to a large extent. Other explanations have gone deeper. In trying to reconstruct the assumptions that researchers had of the political system and their own intervention scope within it, criticism has set in by reexamining the tacit models that served as guidelines for the interaction between social scientists and policy-makers. Best known among these models are the enlightenment model, the engineering model and the dispositional model.[2] The gist of these efforts at reconstruction is clear: the political process was endowed by social scientists with a higher degree of rationality than it actually possessed. Finally, a bit of consolation was derived by pointing out that most of the uses to which social science research has actually been put was of a rather indirect nature. Like in the field of technological innovation, the direct uses are the exception rather than the rule (according to estimates for the field of technological innovation, they do not amount to more than 5 percent).

Changes on the Political Agenda

Thus, it seems that amid all the other structural transformations that can be witnessed in our societies today there is also a new chapter to be opened in the dialogue – which remains a difficult dialogue and one susceptible to interruptions – between the two sides. There is a number of changes to warrant this conclusion.

First of all, familiar boundaries have become blurred. Users or potential users undoubtedly have multiplied, thus enlargening the circle to which social science research is addressed. New types of 'societally relevant knowledge' have appeared, the latest branch of which perhaps is what colleagues and I have termed "service-oriented research," i. e. research carried out in order to provide social groups or individuals with sources of information and knowledge that they can use within a wider political context, but which usually are unable to generate themselves.[3] Most important, however, is the political agenda that has changed rather drastically. Let me mention but a few of these changes.

● The end of economic prosperity: was it but a short interval, an exception in the post-war development?

- The degradation of the environment: this issue has probably led to the shifting of priorities in political agendas everywhere. New forms of political participation have emerged which have spilled over into other policy domains as well.
- Unemployment and the prospect of scarcity of work in the future: this has become not only one of the foremost political worries, but has also led to new types of questions for research. Conflicts of distributions that were thought to belong to the past have become a possibility again, alleviating measures are urgently demanded together with new societal visions, for dealing with the future.
- Science and technology as shaping social relations and life in our societies: after the mechanization of physical labour and mastering the material form of the environment, the immaterial realm is being shaped now. New forms of division of head-labour or of mind-labour are under way and the social impact of scientific and technological developments is so pervasive that social science researchers have to take up the challenge.

One of my hypotheses is that there exists an interdependence between societal developments and the form that research takes, i. e. the kind of research questions that are posed, the methods that are chosen and the practical application that is being aimed at. Obviously, the changes on the political agenda just mentioned have deep repercussions for the social sciences. Their response, in the end, will also determine whether the social sciences will be able to retain their place in the political arena and whether their voice will be heard. Owing to the complexities of the issues involved, boundary-transcendent answers, if not outrightly synergistic approaches, are demanded.

Theoretical Queries

If users, societally relevant knowledge and the issues on the political agenda have changed rather drastically, what does our collective response look like? Obviously, there are two sides to it. One is the theoretical interpretation of on-going societal processes, of phenomena that can be observed now. They carry their own ambivalent meaning which can be unravelled and put on a firmer theoretical ground in a slow process only.

One of the most crucial structural characteristics of changes taking place today, especially in the field of social policy, health and urban development, but equally in the technological field, pertains to the breaking up of old units of analysis. New mixes of social formations appear, decentralization has set in while centralization seems to proceed on to a higher level. Old boundaries are no longer valid and new mixtures, e. g. in the field of welfare, appear that only a few decades ago were entirely unthinkable. It is likely that we are witnessing another leap of complexity when decentralization and the formation of new, smaller units give rise to a re-arrangement of interdependencies. This is not the place to go into details, but I am convinced that I am referring to an experience that today is recurrent in the work of most social science researchers. The old dichotomies of class, of types of society, even the simple schemes of domination and subordination are rapidly losing ground. To be quite clear: societies are still deeply marked by inequalities, power still is a central variable, but a new form of ambivalence in interpreting what is going on has appeared. What to make of the forms of self-help that have come into existence and that are celebrated as the solution to an ailing branch of state-provided social services? How to interpret the changing relationship between men and women, with men (although a tiny minority only) slowly moving into a caring role? What to make of the role assumed by professionals and of their changing profile towards clients and patients who demand greater accountability and control? How to judge in advance the merits of proposals, such as that of a basic, state-guaranteed income which – albeit for different reasons – is being advocated for in practically all political and ideological camps?

In these cases, and in many other ones, it is not easy to state where the progressive elements are of what is merely an old form of exerting social control in a new guise. History teaches us that some groups are winners and some are losers; some have disappeared completely and new, ascendant groups have taken their place. We can be certain that the present transformations that occur with regard to the place of work in society will bring about further deep societal changes. But who is doomed to obsolescence and where are the carriers of the post-industrial theodicy? Social scientists, as Pierre Bourdieu recently pointed out, form part of the process of categorizing, of assigning official identity to social groups which only come into social and political being by virtue of someone speaking in their name or addressing them in this capacity.[4]

There is a lot here for more theoretical reflection which, as the last example shows, has far-reaching consequences for political action. The intermediate structure through which the theoretical work of social scientists (which, at the same time, is immensely political) is translated into a more directly usable form is, of course, the organization of social science research.

Towards Social Science Pluralism

The old days, in which clear-cut interest prevailed and social scientists could choose to work for the administration or the administrative-political establishment, or to join the opposition on the outside, are over. In the first case, there

is a long history from the early needs of statistical observations and fact-gathering of the absolutistic state to the sophisticated latest differentiation and use of research for the needs within the bureaucracies, be it as staff units, as monitoring service units where research needs were closely defined as coinciding with those of the administration and the politically established powers, or as outside contractors. On the other side, action and advocacy research, and the openly politically engaged social scientists who produced them, presented a case for the oppressed and saw themselves as spearheads of social movemenets that would sweep away the existing and hated political arrangements. Such a sharp division, however, was always mitigated in practice. Go-betweens played a major part in reform efforts and academia as well as enlightened scholarship provided a more or less autonomous haven for liberal reform-minded researchers. Yet, in terms of overall interests that guided research, the options seemed more clear-cut than today, since a third party arose that also claimed its share of attention and allegiance from social science research. This third party is represented by ordinary citizens who have a better educational background and who are politically more aware and sensitive. They have raised their voices in new policy arenas, such as the environmental field, where they demand new rights of participation. In a similar development, throughout the health and welfare fields, social movements, citizens' initiatives and local action groups have sprung up, demanding that their everyday concerns are also attended to.
Through the conflicts and controversies that have accompanied the rise of these movements and groups, they learned very quickly that a scientific expertise was one resource among others in the political struggle. Thus, a new function for social science research emerged which makes itself felt by adding a more pluralistic view of the complex social reality. Between the former positivistic ideal of 'objectivity' and 'fact' and the corresponding celebration of the subjective point of view, a more differentiated, reflexive and mediating social science has emerged the task of which i. a. is to help explain the plurality of social locations, interests, standpoints, ideologies and strategies of the social actors. Users and producers of this type of research thus have themselves become much more differentiated and an important step forward has been made towards a social science able to function as the institutionalized reflection of a society on itself.

Lessons for the Organization of Social Science Research

What does this mean for our topic, how can the difficult dialogue be improved? I am very explicit about the consequences to be derived from the analysis presented above. New forms of organizing research are needed, viz. more complex forms with a firm institutionalized basis and greater acknowledgement of the interdependencies that exist between the users and producers of social science research. The external challenges, especially the changes on the political agenda, but of course also the actual changes of social transformation, need to be mutated into new organization structures.

First, it has to be recognized that a need exists for more long-term studies with regard to observing, analyzing and interpreting long-term societal developments. While this does not do away with short-term research, the latter has to be fitted into more long-term studies. One of the lessons that can be drawn from the past decade consists in acknowledging the need for intellectual and political autonomy. Too close ties to the political powers of the day turned out to be a disservice to all parties involved and more often than that, tended to be detrimental to the quality of research.

Secondly, I would like to encourage more experimentation with new mixtures of research organization and of coupling the various stages of the research process rather than cutting it up into arbitrary organizational units. Thus, it could be made a point, for instance, that certain types of more scholarly-oriented research be handed over at a later stage to more action-oriented research. Its task would consist in finding ways of implementing recommendations or in setting up accompanying forms of consultative units servicing either the bureaucracies or other intended users, or both. The typology of different kinds of research at present prevailing ought to give way to a clear conception of research as an ongoing process with some division of intellectual and organization labour between the different phases.

Thirdly, there ought to prevail a more realistic assessment of the role of conflict surrounding or following research. Conflict is to be seen as part of the design and outcome of research that focuses on topics which in themselves are controversial in the political setting. While taking account of this fact, we however feel that arenas for negotiations ought to be set up at the same time. Social reality is negotiated and negotiations to this effect ought to deliberately include an anticipation of what the various actors will make of the strategies at their disposal.

Certainly, not all of these – and other – goals can be achieved if we continue to remain as individualistic in our production of research as we have been so far. In another context I used the example of a football team: even if all of us feel to be good individual players, we need an infrastructure and institutionalized conditions if we ever wish to set up and play as a team. They include the opportunities to practise regularly and not only on Sunday afternoons when politicians – in a very short notice – approach us for producing research bearing on the topic of the day. Another observation is that the (intellectual) star player is in the wrong game; there has to be a well developed understanding of our different –

and complementary – functions as members of a team in which the interplay, by definition, transcends what any individual can perform alone. Only if these conditions are met, social science research will – collectively – be able to perform adequately for an audience called, for lack of a better term, society.

References

1 The desire to be useful to the Prince was reciprocated (or preceded?) by the Prince's desire to receive only "useful" advice. When her son Leopold took office in Tuscany, Empress Maria Theresia suggested to him in 1765 to promote the sciences, "mais celles, qui sont utiles".
2 Björn Wittrock, "Policy Analysis and Policy Making: Towards a Dispositional Model of the University/Government Interface," University of Stockholm, Group for the Study of Higher Education and Research Policy, Report No. 29, October 1983, and "Beyond Organizational Design: Contextuality and the Political Theory of Public Policy," ibid., Report No. 31, February 1984.
3 Björn Wittrock, Peter de Leon and Helga Nowotny, "Choosing Futures; The Evaluation of the Swedish Secretariat for Futures Studies," Stockholm, Forskeningsradnemden 1984.
4 Pierre Bourdieu, "The Social Space and the Genesis of Groups," Social Science Information, vol. 24, no. 2, June 1985.

Marienthal and After
Local Historicity and the Road to Policy Relevance

Local Historicity: In-between-ness and the Distinct Contribution of Austrian Social Science

Science, as we have come to understand in the last decade, is neither ideology-free nor gender-neutral. Nor is it, despite the claims to universalism, free from national constraints, as both the history of science during war times and the postwar history of national science policies convincingly demonstrates. In fact, the 'national components of scientific inquiry' are just beginning to surface (Jamison, forthcoming). Not only have military and economic interests played a leading role in promoting or neglecting certain fields of research, but national traditions, blending cognitive, organizational, and cultural features in a unique mixture, have provided powerful channels for the growth patterns of knowledge production. What holds for the natural sciences has never been doubted for their younger sisters, the social sciences. Too undisguised has been the connection between military and economic interests of European rulers from the eighteenth century onward, with the birth of statistics, demography, and political economy; too intimate has been the relation between the constant threat of disruption triggered off by the process of industrialization and sociology's equally perennial concern with the problem of social order. But while a comprehensive history of national styles and components in scientific development remains to be written, this article wishes nevertheless to address itself to the entangled relationship between local and global historicity, illustrated by the case of Austrian social science in the interwar years. An attempt will be made to show how national contributions – conceived in a specific local political and economic environment – can form an innovative part of an international current of ideas and institutional trends. Furthermore, the general direction that the development of the social sciences took will be described in terms of a shift from politically relevant to policy relevant research; implying a concomitant rise of institutional structures articulating what policy relevance is about. Finally, questions will be raised about the often taken for granted presumed universalism in the social sciences which may turn out to be nothing but a cultural hegemony in disguise, linked to the dominance of a particular country that has taken the lead in shaping what passes

Reprinted with permission from Knowledge, vol. 5, n° 2, Sage Publications, Dec. 1983, 169–192.

for policy relevance through the structuring of its own policymaking institutions and the demands of its technological elite. Related to these questions, we will also ask whether the supposed universalism tends to flourish in times of economic affluence, while in times of recession a contraction and reorientation towards national issues and institutional traditions takes place.

A retrospective second look at Austria – a unique approach? – has the merit that one can follow the evasive shifts of the Zeitgeist while the shadows of history grow longer. What was once a great European power has become a small democratic nation state, situated at the periphery of Eastern and Western Europe. A once-distinctive national culture, in itself heterogeneous and conflict ridden, has joined the choir of other national minicultures after having learned the hard way how to manage conflict. According to Torrence (1981), the development of sociology in Austria proceeded along an original path, but was twice thwarted: its first contribution took place in the peripheral isolation of the provinces of the vast Austro-Hungarian Empire, while its second innovatory contribution, under the banner of Austro-Marxism, came to a brutal end when the clerico-authoritarian regime seized power (Knoll et al., 1981; Rosenmayer and Höllinger, 1969). What followd after World War II is, according to Torrence, merely the 'blossom of an artificial seed', an import from the other side of the Atlantic, even though the transfer came about under the sponsorship of a former Austrian, Paul F. Lazarsfeld. These claims have some merit, but they reinforce the stereotype of 'the science that nobody wanted', while I will try to show that the Austrian contribution corresponded to certain stages in the overall societal development and its institutional framework.

What is undoubtedly the case is the strong focus of the early sociological tradition, associated with the names of Gumplowicz, Ratzenhofer, Ehrlich and others, upon some of the most salient problems of the old Empire – foremost its ethnic problems, to which problems of class were added later and – as behooves a state with a strong bureaucracy and a well-articulated legal system – the development of a 'sociological method of jurisprudence', which laid the foundation for a sociology of law which indeed came to an end in the 1930s. None of the early sociologists, however, as Torrence makes clear, stood a chance, in their pursuit of a sociological pluralism, against the centralist conservative-liberal establishment of the Empire. A politically tightly controlled university system and a very modest degree of sympathy of the Court, even for the natural sciences (Broda, 1981), added obstacles to the establishment of a discipline which was far from being a professionalized field or of showing signs of social utility. It needed the downfall of the Empire and the breaking up of its ethnic diversity which had been unable to overcome their contradictions, to create that special and dense climate of a metropolitan setting which was to become 'the Vienna of the time between the wars'. This in-between-ness is not only a chronological one. Vienna and Austria at that period were literally in-between different cultural identities, born between a suddenly truncated nation state, plagued by doubts about its capacity to survive and the bourgeois nostalgia for the glories of the past which lingered on for quite some time. It was also politically in-between an increasingly polarized political spectrum which stretched irreconcilably from left to right until the situation exploded in a civil war. In-between-ness existed also in spatial terms: amidst the provinces dominated by conservative political majorities. 'Red Vienna' with its clear socialist majority stuck out like a sore thumb, a constant challenge and persistent threat, alive with its pressing housing and other social problems and the vigorous attempts of the municipality to reform housing and welfare policy, the educational system and the city's finances. In-between-ness was even more a pronounced characteristic of the socialist movement. Torn between its radical rhetoric, the constant talk about the impending revolution and its pragmatic, down-to-earth reformist strategies, it was caught in what proved to be a creative cultural tension, albeit one with disastrous political consequences (Leser, 1981).

Austro-Marxism – a term that was originally intended as insult – came to denote the theoretical work of a group of young academic Marxists, which was a regional variant of international Marxism characterized by its attempts to develop an alternative between reformism and bolshevism. It also became synonymous with the day-to-day tactics and political practice which the socialist movement deployed in building up its organizational network for restructuring the whole cultural life sphere of its members (Leser, 1981). The thrust of these efforts was directed, as is well known, towards creating parallel structures – literally a counter-world against the bourgeois lifestyles. Multiple networks and organizational ties were to encapsulate the entire life of party members (amounting to a third of the Viennese population) in their leisure and education, from childhood to funeral services. Lacking the political clout to effect changes in the sphere of production, the effort to effect any changes at all turned upon the building of an elaborate organizational network. In their temporal in-between-ness, the leaders of the movement saw this concrete, yet principled, takeover of the cultural sphere as an anticipation of a historical future yet to come. The blocked, politically not-feasible restructuring of work conditions and control mechanism was compensated by creating a borrowed future – in the life sphere outside work, which however, led to isolation and a tragic underestimation of the real dangers that were to come from the political outside.

It is in this general and unique climate of in-between-ness where cultural efflorescence and innovations took place. The confrontation with Austro-Marxism and its own temporal in-between-ness is the underlying thread colouring also the rhetorical and institutional developments in the social sciences of that time. In the following, I will examine three examples: the debate between the Viennese

School of Economics and its Austro-Marxist counterpart; the strand of thought directed towards social planning and a rationalistic world conception, exemplified in the life and work of Otto Neurath; and finally, the beginnings of applied social research, in its Austrian form an anticipation of an institutional model, which Paul F. Lazarsfeld was to realize later in the American context. Although I will draw upon the work of individuals, this should, nevertheless, be interpreted as reconstruction of lines of development that transcend individual biographies.

Planned Economies, Scientific World Conceptions, and the Unemployed of Marienthal

Properly speaking, the fundamental works of the Viennese School of Marginal Utility, with their emphasis upon the subjective side of economic value theory, fell still into the period of the old Austro-Hungarian Empire, namely 1870–1890. Their continuing influence into the 1930s and – in a certain sense – even their survival well into the 1950s, justifies it, however, to treat the ever renewed confrontation with Marxist economic theory, in its specific Austrian variant, as an important contribution of the time we are dealing with (März, 1981). The two schools of economics literally lived in different problem worlds, but drew at least implicitly upon each others' work in their heated polemics. The acuteness of economic problems – above all the misery of massive unemployment, the social consequences of inflation, and the concentration of bank and industrial capital – added fuel to the debate about the feasibility, urgency, and questioned or asserted efficiency of a centrally planned and administered economy. Both schools wrote with an engagement and solid craftmanship which was never before and, unfortunately, never afterwards reached again (März, 1981). But although the actual problems were acute, the debate remained essentially a theoretical one, not only was economics a well-established academic discipline with a high reputation which set the standards of the debate, not only were the main proponents of Marxist economic theory well-educated 'academic Marxists', but the politcal situation was far too unstable to allow anything like 'putting economic theory to a test', as we are witnessing today with Reaganomics. Schumpeter's brief interlude as Minister of Finance in Renner's cabinet illustrates this very well (Flos et al., 1983). It was overshadowed by personal and political conflicts with his colleagues and ended with his dubious involvement in a deal in which shares of one of the largest Austrian industrial enterprises were sold to an Italian group – which put an end to any serious nationalization plans of the government. Thus, while the practical input remained low, the confrontation of two strongly opposed theoretical and practical programmes of economic

134

dogma and policy received its special flavour and high standards from the academic context and the political brisance of the main issues alike.

That the theoretical positions of Austro-Marxism were far from homogeneous can be seen when comparing such diverse intellectual orientations as those of Otto Bauer, Max Adler, and Otto Neurath. What united them was their streak of activism, but the 'practical utopia' reached its fullest and most creative expression in Otto Neurath's life and work. His guiding vision was what he termed the scientific world conception. In a book of the same title, he blended Marx, Comte, and behaviourism into a kind of empirical sociology of the future which should provide the rational guidelines for all forms of personal and public life, for education as much as for architecture, for social life as well as economic life (Neurath, 1979). Sociology should become a synthesis of history and economics and since it was concerned with human behaviour, its scientific basis demanded to view life as temporal-spatial processes in which figurations of humans and their environment occurred. Since only science was capable of providing a unified system of statements with the help of which controllable forecasts about temporal-spatial processes could be made, Marxism became for him the scientific tool par excellence for planning history, since he saw in it the best example of a strictly scientific and unmetaphysical physicalistic sociology.

Neurath combined a unique set of interest and skills: he was an early and prominent theoretician of the so-called war economics, a term which he apparently coined himself, a social technician and expert for nationalization questions, a leading member of the Vienna circle, and the founder of a popular version of visual statistics.[1] His early pre-occupation with the conditions under which wars influenced population development, supply and demand, problems of monetary flow, organization of food procuration, and their impact upon money and banking made him realize that during war times the usual allocation mechanisms of the market were replaced by administrative mechanisms, while money and monetary flow became a kind of macro 'natural goods' economy. Since he observed cases in which the general welfare of the population increased as a result of war economy, Neurath drew the conclusion that one had to transfer the planning features of his economy into peace time. The instruments which were created in war had to be utilized for a conscious life planning in peace.

His fascination with the practice of planning, derived from his theoretical work, made him a much sought after-expert on nationalization questions, although his political fate as expert was bound up with the political fate of those whom he sought to advise (the Räterepublik in Bavaria). It also led him to what he considered to be the task of the new strategic age; to plan society, since everything should become transparent and controllable. He was a social optimist par excellence, a proponent of socio-technics and the possibilities for steering social progress with technical means; culminating in the conviction

that a significant part of our life order can be formed purposefully, that espe-
cially consumption and production can be determined quantitatively and con-
trolled, even though we can not yet control with socio-technics ethics and cus-
toms, religion and love or do not wish to do so.

Comparing the performance of an economy in war and peace time had led him to believe in the greater efficiency of a planned economy; by extending his planning model into one of scientific rationality the 'scientific world conception' came to be the key for understanding the world and for controlling it. Socialism and the democratic planning of material production were for him the 'institutionalization of strategic-technical thinking for the realization of human aims', especially in creating a more efficient and human economic order. The founding of various museums for the purpose of educating the masses in the basics of economic and social thought can also be interpreted as an education in cognitive skills deemed necessary for participation in the democratic planning process. But above all, it was the scientific world conception and the elaboration of its epistemological foundations through logical empiricism which should become the scientific basis for 'serving life', including the everyday life of ordinary people. When Otto Neurath died in 1945, his life had been full of difficulties, but also intensity. His involvement with nationalization issues brought him before a political trial for high treason and the academic establishment remained closed to one of the foremost protagonists of the scientific world conception. For a significant part of his life, he was a political refugee in exile, well acquainted with poverty. But he never seemed to have lost his belief into the possibility of designing effective strategies for bringing about collective happiness and he remained an adherent of 'utopistics as science': such a science would have to construct not only one utopia, but a whole set of them, comparing different designs of social order.

In her memoirs and reflections on the origins of applied social research in Austria, Marie Jahoda paints a vivid picture of the pervasive influence that Austro-Marxism and the social democratic movement had upon young intellectuals on the left: Austro-Marxism was not just a theory, but a world view. She singles out three aspects of significance for the beginning of social research: the belief into the possibility of a humanitarian, democratic socialism; the emphasis upon the present with its rich cultural transformations; and finally, the educational function of Austro-Marxism – i. e., its attempt to educate the workers and to render visible the historical dimension (Jahoda, 1981).

Not surprisingly, given the high priority attached to reform of the educational system in and outside school as well as the dominance of the socialist municipality in this domain, one of the early crystallization points became the psychological institute of Karl and Charlotte Bühler. The other focal point, but apparently more devoted to 'social book-keeping' than social science research, was the Chamber of Labour, where a number of studies on the living conditions of deprived groups had been undertaken with the explicit aim to improve their condition. In Marie Jahoda's account, the naivité of the young researchers who wanted to undertake research in order to draw consequences for political action and their carelessness vis- à-vis established disciplinary boundaries gives them the fresh approach of highly motivated amateurs.

It was around the Bühler's institute of psychology – a kind of third way between Freud and Adler – where the institutional genesis of applied social research took place. The Bühlers were internationally known and kept an open house for visitors from abroad. Much of the work pursued at their institute was done on a contract basis, most of it with the municipality of Vienna. When Lazarsfeld was able to found his "Wirtschaftspsychologische Forschungsstelle" – an institutional innovation which he considered among his greatest achievements – a new kind of organizational tie was established between what was later to become a world wide practical model of contract research based on the relationship between sponsor, client, and researcher (Lazarsfeld and Reitz, 1975). Lazarsfeld's famous adoption of a well-known quote of the day 'that a coming revolution needs economics (Marx); a successful revolution needs engineers (Russia), while a defeated revolution has made social psychologists out of us (Vienna)' contains in a self-ironizing way already the tension which would later become apparent with the mounting pressure of commercialization upon this model. But in the early 1930s the sociographic concerns, as the authors called them, were still very much tied to the social democratic movement; out of which arose the most important work of the group, the Unemployed of Marienthal (März, 1983). What was studied were not the individual unemployed, but the whole unemployed village, the changes and repercussions of prolonged unemployment on lifestyles, perceptions and political morale. Its general conclusion was verified not much later by the success which Hitler scored also among social democratic workers – namely, that prolonged unemployment leads to resignation and not towards revolution. Highly original in its methodological conception, which was ironically partly due to the lack of a clearly formulated research design,

Marienthal grew out of our will to know, out of our contacts with the unemployed in the political movement, out of the numerous improvisations and out of a work team, the roots of which lay in the youth movement in which neither a formal division of labour existed, nor a systematic accounting procedure; out of our world view and out out of an intellectual discipline, which we had slowly acquired through our university studies through market research. Methods resulted from the concentration upon a problem, not for their own sake. (Jahoda, 1981)

The importance of seizing upon a theme at a period where the empirical phenomenon under investigation was widespread, its political brisance, a highly developed sensitivity towards structural factors combined with an unusual mixture of qualitative and quantitative methodological principles, point towards what could have become the institutionalization of a new school of empirical research, which would have combined scientific discipline and methodological innovation with reformist concerns of 'making the invisible visible.' The reasons this did not take place in Austria are known: Marienthal was published in 1933; the greatest part of the unsold copies were burned shortly afterwards.

Local Historicity and Global Developments: The Missing Link

What is striking about the three cases examined in the context of local historicity is the degree to which they partook in what was equally an international movement of ideas, concerns, and tendencies. As we have already seen, the continuing influence of the Viennese School of Marginal Utility was renewed by their equally continued confrontation with Karl Marx's value theory and the interpretation that the young Austro-Marxist economists made of emerging socioeconomic phenomena like monopoly capitalism, finance capital, imperialism and the economic aspects of the ethnic conflicts. The influence of the Viennese School survived the actual life span of its founding members, and also of Austria – 'their home country which had become narrow and poor' (März, 1981). At the end of the war, only von Wieser was still alive; among the next generation Ludwig von Mises, Hayek, Haberler, Morgenstern, and Machlup were soon to leave Austria in order to teach in more internationally minded universities of the North American continent. On the other side of the debate, Hilferding, perhaps the most original economist among the Austro-Marxists, had left Austria early for Germany, while Otto Bauer and Karl Renner became more preoccupied with their political work and legal studies, respectively. In an essay written in 1927, Joseph Schumpeter concludes that the influence of the Viennese School is still dominant, and Walter Schiff's 'Planwirtschaft und ihre ökonomischen Probleme', published in 1932, testifies to the continued salience of the nationalization issues. According to März, Hayek's 'Weg zur Knechtschaft', which appeared after World War II, is but a continuation of von Mises's earlier work – part of the unrelenting debate about the fundamental irrationality of a planned economy functioning according to socialist principles. Thus, it is not only a fact that the main points of the controversy touched the essential theoretical and political issues of economic policy, but it is also due to the highly articulated and well-worked out positions of Bauer, von Moses, Schumpeter, and Oskar Lange that the debate reached high international standards and scope, well beyond the tragic fate of Austria at that time.

Turning towards the work of the Vienna Circle, we can observe in a similar vein that the issues and the intellectual movement they were soon to engender lasted well beyond the actual survival of individual members. Logical empiricism became a worldwide philosophical movement already in the 1930s, with a string of international congresses organized around themes like those of the unity of the sciences or the fundamentals in mathematics. While the Vienna Circle has long ceased to exist in Vienna, its individual members having fallen victim to the authoritarian regime or fascism, assassination, death, or emigration, the movement spread worldwide and dominated philosophy for decades to come. Although the concerns with planning on a scientific basis took a personal form of expression in the work and life of Otto Neurath, the main thrust behind the scientific world conception, the epistomological foundation of a planned and highly rational design of 'forms of life', have to be seen as part of a more widespread antimetaphysical stance which was to be found in England as well as in the United States. The fascination that the Sovjet Union exerted upon intellectuals of leftist or even Fabian persuasion at that time is well known – new fuel was added to the political imagination and utopian visions on an almost day- to-day basis.

Likewise, applied social research was of course not unique to Austria and was far from originating there. As Hans Zeisel makes clear in the methodological appendix to the Marienthal study, survey methods were practised long before in England and in France, and Lazarsfeld would later show that empirical social research had an extensive history of its own.

The original contribution made by Lazarsfeld and his colleagues consisted in inventing a new organizational model, which, after its successful implantation in the United States, was re-imported to Europe in the late 1950s and 1960s. Jahoda remarks that the international atmosphere which prevailed at the Bühler's institute taught her and her colleagues that 'national specifics in research can only be legitimate, if they can be justified before an international forum', but this applies also to institutions: Lazarsfeld's Wirtschaftspsychologische Forschungsstelle became, despite its humble local origins, part of a worldwide movement of institutionalizing applied social research.

It is therefore not primarily due to the emigration of some of the leading figures or their students to various parts of the world and their successful activity outside Austria that we can explain the international intellectual scope of the cases we have analyzed. The point is rather that the Austrian contributions were part of a wider international movement of ideas, innovations, and institutional strategies that had their outposts also elsewhere, but to which the Austrian case added highly specific and unique features. This international movement lived on and incorporated the Austrian contributions, long after Austria had ceased to contribute anything. Although personal fates were often tragic, it probably aided the diffusion of the Austrian contribution that some of its proponents were able to amalgamate the more local features of origin with the new environment into which they had moved. But this was not the cause.

We therefore have to return to the question, how it is possible that a small country in central Europe, one that 'had become narrow and poor', was part of a worldwide stream of global developments? More precisely, what was the international character of the problems and underlying issues which were addressed also in the Austrian scientific vernacular?

The Road Towards Usefulness: In-between-ness and Pre-institutionalization of Policy Relevance

The period we have been looking at cannot be understood properly without according Austro-Marxism its place and delineating its influence. Compared to other 'critical' theory systems like the Frankfurt school, Austro-Marxism was far more practice-oriented and, in a certain sense, was able to create the conditions out of which further theoretical work could evolve. The new group of what were for the first time 'academic Marxists' – who would draw Trotsky's ridicule[2] – were committed intellectuals, tied to each other through friendship and personal ties, and strongly motivated under the early leadership of Grünberg, to fight against the prevailing metaphysical spirit and what were considered obscurantist traditions. They were taught that 'categories and conditions . . . always belong together . . . to the present conditions their historical antecedents have to be added'. While theorizing was undoubtedly directed towards practice, its actual function was largely to serve as legitimizing political practice, until the tragic clash with an unruly political reality occurred. The more political developments narrowed the degrees of freedom for theory to guide practice, the stronger the urge became to use science in order to change reality. The scientistic streak is most obvious within the Vienna Circle, but divided therein between a kind of political abstinence and Otto Neurath's utopian vision to construct human happiness with the help of a scientifically derived 'rational practice of life'. The in-between-ness of Austro-Marxism expresses itself once more: since there were few real political issues which could be influenced directly, theoretical work and scientific inquiry took the form of being *about* a future social order, which was seen as having begun *already*: nationalization, for instance, was not only debated in theory, but steps of how to bring it about and make it function were devised; adult education was one of the most important vehicles in bringing about a new and more humane future, into which Otto Neurath's visual statistics and pioneer efforts in the popularization of economics fitted beautifully, while Lazarsfeld's early research on the occupational choice of proletarian youth could be seen as leading up to an expanding network of counselling services provided by the City of Vienna. Yet, in all these cases, the vision of a future was necessarily coupled with its historical past which Austro-Marxism could not fail but to reanalyze. National and even local problems were the concrete exemplars, the 'Zuständlichkeiten' in the sense of Grünberg, that were embedded not only in a theoretical framework, but above all in a wider historical *and* future-oriented perspective. It was not a hallow pseudo-universalism, as would become the case later when under the banner of a thinly disguised ethnocentric modernization theory world models would be stimulated with neither past or present; nor was it the degenerate version of much of American commercialized applied social research, under which any problem would become immediatized under the pretext of policy-relevance.

In the time between the wars, policy-relevance was not yet born as a category, nor was it politically feasible. Of course, the theoretical tenet of Austro-Marxism, as of other world views, was that theory should guide action. The Vienna Circle even went so far as to postulate that the scientific world view should be made the fundament of any kind of individual or collective action. But these were not criteria of policy relevance as they would be understood later. Paradoxically, the state of in-between-ness provided a secure anchorage for the present, since it was thought to be known into what kind of future the past was leading. Practice orientation meant to move ideas in a somewhat diffuse political and personal network, mediated by chance encounters and indirect influences, which were nevertheless expected to exert an impact in one direction. There was no 'decision-maker' yet to address, since no institutional framework existed that was geared towards soliciting, screening, and utilizing policy advice. The structures that existed were still in the highly personalized stage of a movement in which intellectuals played a leading yet relatively undifferentiated role. However, pre-institutionalization as I have called it, of the modern relationship between applied social science and politics was already under way.

While pre-institutionalization in the field of economic policy followed the established pattern of interelite advice and consultations with a number of finance ministers already under the old Empire coming from the leading ranks in academic economics, the precariousness of a fascinating new venture which in the end failed to materialize, is represented at its best with all its contradictions in the work and the life of Otto Neurath. His attempt to bridge popular aspirations and popularization of the sciences at the level of adult education with the pursuit of epistemological inquiries of a unified science at the highest level of academic standards contains a grandiose vision of institutional structures to create. They should allow of the big jump forward on the road to modernity which in his view meant socio-technics and the commitment of science to a rational politics in the service of human betterment. The turbulence of the times and of his own life biography were too great to offer more than a chance for institutionalization, but we may wonder why such chances are lost or diverted into far less inspired, yet apparently more stable institutional channels. Otto Neurath was one of the great practical utopians, whose vision of uniting practice and theory guide, and often misguide them, but who serves as active ferment in the overall changes of institutional structures out of which innovations emerge. His ideas about rational politics, still inspired by his version of socialism, were to reappear after his death in the form of the comparatively flat and uninspiring policy sciences. Undoubtedly he would have approved of the use of computer models for forecasting and technological assessment, but he would have been the first one to demand that the public should have access to the new skills and be fully incorporated into planning their own future. In proposing his 'Lebenlagen-Kataster', he anticipated by almost half-a-century life style research and social accounting procedures; but who among the social indicator adherents would dare today to call statistics the fundamental of human sympathy or see in the proletariat the avant-garde of science without metaphysics?

The hero in my story of pre-institutionalization is of course Paul F. Lazarsfeld, but he is a flawed hero. His and his collaborates' contribution came at a time when policy-prone institutions – the Municipality of Vienna, but also the Austrian Broadcasting Corporation and industrial firms, began to be interested in new consumption patterns which were needed for mass produced goods and services. It is significant that his 'Wirtschaftspsychologische Forschungsstelle' was not part of the existing academic framework and had no chance of being integrated there. But the marginal position of the new applied social research proved to be surprisingly strong, as it turned out when the institutional innovation was later successfully transplanted into the American environment.

Much has been written on the strength or weakness of Lazarsfeld's political engagement or – to put it into less personalistic terms – it has been asked what kind of socialism this was that enabled him to make his famous remark about the 'methodological equivalence of socialist voting behaviour and the buying of soap' (Gitlin, 1978). In my account, the underlying question is about the kind of political engagement as it was practised under the influence of Austro-Marxism and the kind of highly professionalized politically aseptic user-client relationship that claimed policy relevance later. It amounts to letting oneself become confused by superficial differences in context and historical idiosyncrasies if one were to separate what is really a similar evolutionary mechanism in the emerging role of social science research in a changing societal structure. The rise of the social sciences in Europe was tied to the opportunities and vagaries of an essentially moderate political reformism. As Pollack has argued, reformist practice prevailed in the actual policies of all socialist-democratic parties in Europe since 1890, despite their radical rhetoric (Pollack, 1981). The visible transition from the concept of revolution to that of the 'social question' and social policy issues could not but affect a whole generation of party leaders and union functionaries who were already more accustomed to administration than agitation and who realized early the importance of social science information for their strategies, including the probing into the interest and needs of what happened at the grass-root level.

In the U.S. context, the broader societal changes were also clearly reflected in the rise and decline of 'schools' in sociology that played the leading role in the discipline at different times in history (Coleman, n. d.). The shift can be illustrated from the the preoccupation with problems at the local level – for which the Chicago School was exemplary – directed in its proto-policy relevance at a diffuse audience of City Hall politicians, journalists, reform-minded honoric citizens, and philanthropists, to an entirely new set of problems which arose from structural changes at the national level. Selling to national markets, from breakfast cereals to radio programmes, and forging new communication structures on the political level akin to the commercial became the hallmark of the new corporate actors to whom sociology – epitomized by the Columbia School – would now address itself to. The earlier locality-specific interests were replaced by new social relations between the social scientists and the large corporate actors, who themselves needed and thrived on a national network of economic production, marketing, and control (Gitlin, 1978). Thus, social science research with its shifting yet guiding set of problems is neither immutable nor can it simply be grasped in a one-dimensional historical development. Rather, it changes its nature in response to and anticipation of changes in the societal structure that bring forth new social actors and their power to define a problem agenda, as well as the creation of new institutional arrangements. While the user-client model relationship of a highly professionalized social science research enterprise was clearly absent in Austria in the interwar period, just as institutions geared to the intake of policy advice were either entirely missing or in a nascent stage on the municipal level, the preconditions for the institutionalization of the new kind of research undertaking for which Lazarsfeld stands as a pioneer entrepreneur, were already there – providing the link between local and global usefulness of the social sciences.

The inevitable erosion of whatever illusions about the extent of political commitment might have lingered on and the increasing incorporation of social science research into the 'administrative complex' was certainly facilitated by the salience accorded to methodology. But in order to fully appreciate the extent of the shift from a politically relevant research to policy relevance, we have to look at the real success of Lazarsfeld's institutional innovation – namely, its re-importation to Europe in the 1960s. The hegemony of American sociology after the war, and the victory of empirical sociology in particular which promised crisis management techniques to policymakers through its technical lead facilitated the institutionalization of the kind of applied social research which was initiated in Vienna in the 1930s.

It provided an ideal vehicle with the help of which the intellectual and ideological 'modernization of Europe, the strengthening of political pragmatism and the corresponding weakening of Marxism' played a key role (Pollak, 1981). In the wake of the successful transplantation of empirical sociology into contexts with different national traditions, the history of the discipline had to be rewritten also (Oberschall, 1981). In his survey article for UNESCO, Lazarsfeld speaks of sociology as 'an American science' and it is obvious that he believed that the sponsor-researcher relationship of social research would apply on a global scale. Indeed, as Pollak (1981) put it, the new institutional model offered something for everyone:

In this holy alliance for the benefit of empiricism and applied social science research everyone could remain tied to their political ideology while referring to the increased utility for policy-makers that the new science would yield: for the American businessman the new science promised the improvement of marketing strategies, to the social-democratic politician a weakening of tradition-bound, but obsolete ideologies, to the conservatives the struggle against communist utopia and to all administrators the rationalization of decisions.

The 'substantive implications of methodological procedures' about which Lazarsfeld was fondly and wisely speaking turned out to be wider than he himself anticipated. Methodologies do indeed have substantive implications, but they also have institutional prerequisites and it is within a certain political arrangemnt that they are most centrally in command – and commanded. The hopeful but naive slogan of the Enlightenment – *savoir, pour prévoir, pour pouvoir* – became the crux of the social sciences. In their self-inflicted usefulness, as Schumpeter once called it, they are moving along with those forces in society that want them to be useful – arranging at the same time the scene in such a way that they indeed succeed in proving their usefulness.

Marienthal and After: Is Scientific Mediocrity the Fate of Small Nations?

In Austria, his home country, Lazarsfeld, together with Oskar Morgenstern, became the founding father of the Institute for Advanced Studies which was intended to play a strategic role between Eastern and Western Europe and within Austria in opening the social sciences to formal methods and empirical techniques which the universities were ill-equipped to provide (1958 – 1964). This time, the institutionalization of applied research met with corresponding political institutional structures, certainly in the field of economic policy. Intellectual modernization with corresponding institutional structures in policy-making had finally reached Austria.

The rest of the story of the social sciences in Austria (from 1945 onward) is briefly told: in the reconstruction period of the universities, the social sciences proceeded only slowly, with old vestiges of ideological struggles lingering on for a long time. The main achievements after the war are not to be found so much within the universities, not within any particular discipline or field of research, but lie in the field of institutional innovations. The 'Sozialpartnerschaft', a truly unique institution in the political arena, found its scientific policy analogue, for instance, in the 'Scientific Council for Economic and Social Questions'. This Beirat, founded at a time when economic policy was under assault of how to integrate Austria into the European market and under the internal threat of the coalition breaking apart, has been interpreted as the rise to power of a new technocratic elite, consisting mainly of young economists (Marin, 1982). The unanimous priority attached to a policy of economic growth contained the promise for an almost conflict-free satisfaction of all interest groups which could articulate their needs under the guidance of economic empirical knowledge. The new technocratic elite with its claims to monopoly on this kind of knowledge could draw upon concepts and symbols of programming, planification, and Neo-Keynesianism that were internationally in good standing as part of a modern, concentrated economic policy within a reasonable middle-range orientation.

Other institutional innovations were less spectacular and more indirect in their policy relevant aspects. The funding establishment of a Ministry of Science and Research in 1970 when the present government came to power certainly helped in putting the social sciences on a firmer basis. The theoretical work performed at the Institute for Advanced Studies, especially in the area of economic forecasting, was fully incorporated into an already existing expert network of policy advice. Practically all other institutions in which social science research in one form or another is practised were founded outside the university system with more or less direct ties to ministries or agencies interested in their services. In some instances, ministries started even to expand their internal capacity to initiate, coordinate, and utilize social science research. Characteristically, one of the most important funding agencies in social science research is within the sphere of influence of one of the main proponents of social partnership, the Jubilee Fund of the Austrian National Bank, which is controlled neither by the government nor through the universities, while institutions attached to the university system, like those of the Boltzmann Gesellschaft, are operating on a rather low budget.

It is fair to say, I hope, that social research in Austria has followed patterns as they exist in other small European countries today: it exihibits a relatively low degree of professionalization and exists in an elaborate extra-university structure of usually quite small institutes with a high degree of dependency upon a small number of concentrated financial resources. There are few signs of any formal or explicit science policy for the social sciences (apart from official statements, of course). Yet, social science policy is alive in a highly informal network, operating within institutional structures which follow closely the overall logic of the dominant political structures. With a few exceptions, social science research at the universities remains relatively segregated from the work performed in the more policy-oriented institutes, which also has its political reason.[3] Local problems and relatively short-term policy orientations prevail. Spectacular and internationally renown achievements in the social sciences have become rare. Is this the price to be paid for intellectual modernity? Was Marienthal indeed, as Jahoda has maintained, the result of a unique constellation which has not occurred again in the life work of any of the collaborators of the original study – not to speak metaphorically of the chances for replication in the life of a nation? Is the price to be paid for intellectual modernity – with its high degree of policy relevance and utilization – always scientific mediocrity?

Conclusions

In order to answer this question, we have to look at two different kinds of relationship once more. One is the intricate connection between changes in societal development and the nature of social science research. We cannot start from the naive assumption that the pressing social and economic problems in any society are but the raw material to which an enlightened social science has but to address itself in order to come forth with solutions. Rather, the interplay between themes, conceptual and methodological tools of inquiry, and the problems which underlie them, is a subtle and complex one, mediated through the institutionalized forms which social science research has achieved in a particular time and place. There can be no doubt that in the 1930s a whole gamut of pressing social and economic problems were highly visible. There can also be no doubt that they were taken up and addressed by social science research, but this response still occurred in an institutionally unmediated way, the pre-institutionalization stage of policy relevance. Therefore, a major part of the answers lay in the search *about* the feasibility, efficiency, or rationality of nationalization – for example, about ways how to address the problems of a faltering market economy fully exposed to the forces of the world market. Otto Neurath's vision of social engineering was *about* the proper strategies of coping in a rationalistic way with life, and about means and ways that would enable popular participation in scientifcally derived coping strategies. The Marienthal study, although suggested by Otto Bauer as a theme, was not commissioned by any ministry or party functionary. Rather, it arose out of the enthusiasm of engaged researchers who wanted to effect change by making visible the invisible. It contains a codification of proposed principles for similar studies in which the authors laid down their proposals on what kind of data should be collected according to what principles; it is about the research strategy to be followed in the future.

We have examined a context in which the mediating structures of policy making and research oriented towards it were yet to be created. The sense of urgency and commitment, the premonitions and historical awareness which are so apparent in the works of that time, derives from their spanning the whole gamut from past to future, in which the present is simply a strategic place from which to influence the future. By contrast, present policy-research often lacks past and future orientation. It is fitted into an existing and well-functioning institutional framework, in which the future has been subsumed under the present. Its success, measured in its impact on policy making, derives partly from the monopoly a technocratic elite was able to establish worldwide on 'its' knowledge and to control its empirical basis; but partly it is also a function of the ability to depoliticize problems by turning them into technical ones, which are by definition open to expert advice only. The uniqueness of the Austrian post-1945 institutional innovation lies in the fact that it acknowledged that objective advice is impossible in politically sensitive areas, and of incorporating bipartisan advice in a carefully balanced system of political negotiations. The success of such a strategy hinges, however, upon the prior consensus over which political questions are to be transformed into technical ones, and the maintenance of the consensus.

The other kind of relationship to be examined once more is the link between the national and the international dimension. When I emphasized the truly international scope of the Austrian contributions in the 1930s, this is more than a nostalgic nod. Rather, we have to ask how the affinity to the movement of ideas, themes, and tendencies, which also occurred elsewhere, came about and what were the mechanisms through which the main proponents took their stand in an international debate. Under the hegemony of American social science and empirical research after World War II, reinforced through the advance of computers, a certain pseudo-universalistic tendency was to spread; it was widely believed that the methods of applied social science research would provide the kind of neutral tools that could be put to use anywhere – analogous to the myth about the neutrality of technology. In the technocratic conception of social research, national differences are reduced to insignificant variations in a world system that claims to be neutral and universal in applicability. This myth was sustained in times of economic growth and affluence when many nations underwent a push towards modernizing their elites and decision-making procedures and were able to afford the luxury of importing institutions from abroad that promised to contain the latest advance in scientific methodology. In times of economic recession, however, the supposed universality and neutrality of technocratic knowledge is undergoing a severe test. The problems with which many national economies are confronted – including their social consequences – are how to maintain their advantage in a world economy. If all knowledge to do so would be readily available in a kind of big data bank, ready to be retrieved by whoever wanted to, everyone would be successful! This is blatantly not the case. We can therefore assume that national differences in style and content of social inquiry are in the ascendance once more. Economic and social problems put on a national face, which is only reinforced through international dependencies. If the economic recession continued, social research will become more redirected again towards the peculiarities of its national problems and also reactivate national cultural traditions of inquiry. The slow but perceptible decline of American sociology, for instance, has not only to do with its waning hegemony in the world order. In times of crisis, the Zeitgeist also gets restless and moves on, hovering here and there in order to materialize in unexpected and peripheral places, provided – and this is a difficult condition to meet – that local historicity is bound up again in the right kind of mixture with global developments.

Thus, true universalism – as distinct from pseudo-universalism which only masks the hegemony of a particular national tradition or of a particular group, even if it is spread internationally, is by definition not only pluralistic, but even more important, can only be achieved when being constantly reinvigorated and infused with cultural diversity. While any local or national scientific culture is always in danger of falling into sheer provincialism or obscurantism when it missed to check itself against international standards, the conditions of universalism are much more difficult to meet. They presuppose that cultural diversity is both creative from its local origins and open to be integrated into a global development which does not negate diversity. Science has contained this duality all along and we are challenged once more to put it to creative use; for the benefit of local science and for the benefit of universal science alike.

References

1 Otto Neurath founded in 1924 in Vienna the 'Gesellschafts- und Wirtschaftsmuseum', which he called an 'adult education institute for social enlightenment'. During his exile in the Netherlands, he initiated the International Foundation for the Promotion of Visual Education by the Vienna-Method and the Mundaneum Institute of The Hague.
2 Leon Trotsky, who spent 7 prewar years in Vienna, looked the prominent Austro-marxists over and commented 'In the old imperial, hierarchic, vain and futile Vienna, the academic Marxists would refer to each other with a sort of sensous delight as "Herr Doktor".' 'He went on 'they were knowledgeable, but provincial, philistine, chauvinistic. . . These people prided themselves on being realistic and on being businesslike . . . but despite their ambition, they were possessed of a 'ridiculous mandarin attitude'.
3 According to Bernd Marin, in the beginning of the 1970s, two-thirds of all economic and social scientists in Austria were either conservative in their political affiliation or considered themselves to be 'apolitical'. See Marin, B. (1978) Politische Organisation sozialwissenschaftlicher Forschungsarbeit, Tables 18–21, pp. 195–198. Wien: Braumüller.

Bibliography

Broda, E. (1981) "Naturwissenschaftliche Leistungen im gesellschaftlichen Zusammenhang", in N. Leser (ed.) Das geistige Leben Wiens in der Zwischenkriegszeit. Wien: Bundesverlag.
Coleman, J. (1980) The Structure of Society and the Nature of Social Research. Knowledge, 1 (3) 330–350.
Flos, B., M. Freund, and J. Marton (1983) "Marienthal 1930–1980". J. für Sozialforschung, 23, Heft 1: 137ff.
Gitlin, T. (1978) "Leon Trotsky, my life, 1929." Media Sociology, Theory and Decision 6, 2: 241–242.
Jahoda, M. (1981) "Aus den Anfängen der sozialwissenschaftlichen Forschung in Österreich", in N. Leser (ed.) Das geistige Leben Wiens in der Zwischenkriegszeit. Wien: Bundesverlag.
Jamison, A. (forthcoming) The National Components of Scientific Inquiry.
Knoll, R., G. Majce, H. Weiss, and G. Wieser (1981) "Der österreichische Beitrag zur Soziologie von der Jahrhundertwende bis 1938", in R. König (ed.) Soziologie in Deutschland und Österreich, 1918–1945. Kölner Zeitschrift für Soziologie, Sonderheft 23.
Lazarsfeld, P. F. and G. G. Reitz (1975) An Introduction to Applied Sociology. New York: Elsevier.
Leser, N. (1981) "Austromarxistisches Geistes- und Kulturleben", in N. Leser (ed.) Das geistige Leben Wiens in der Zwischenkriegszeit. Wien: Bundesverlag.
Marin, B. (1982) Die Paritätische Kommission – Aufgeklärter Technokorporatismus in Österreich, Wien.
März, E. (1983) "Joseph A. Schumpeter as Minister of Finance of the First Republic of Austria, March 1919–October 1919", in H. Frisch (ed.) Schumpeterian Economics. New York: Praeger.
März, E. (1981) "Grosse Denker der Nationalökonomie in der Zwischenkriegszeit", in N. Leser (ed.) Das geistige Leben Wiens in der Zwischenkriegszeit. Wien: Bundesverlag.
Neurath, O. (1979) in Rainer Hegselmann (ed.) Wissenschaftliche Auffassung, Sozialismus und Logischer Empirismus. Suhrkamp.
Oberschall, A. R. (1981) "Paul F. Lazarsfeld und die Geschichte der empirischen Sozialforschung", in W. Lepenies (ed.) Geschichte der Soziologie, Band 3. Suhrkamp.
Pollak, M. (1981) "Paul F. Lazarsfeld – Gründer eines multinationalen Wissenschaftskonzerns", in W. Lepenies (ed.) Geschichte der Soziologie, Band 3. Suhrkamp.
Torrence. J. (1981) "Die Entstehung der Soziologie in Österreich, 1985–1935", in W. Lepenies (ed.) Geschichte der Soziologie, Band 3. Suhrkamp.

Heroism, Order and Collective Self-understanding: Images of the Social Sciences

The Social Production of Images

The social sciences are in a peculiarly dual position when posing questions about images of science: they can claim some knowledge about the social production of such images and they are themselves producers like the other sciences. Yet, the knowledge they possess about images does not necessarily improve their performance as producers. Here, as in a more general sense, the social sciences are constrained in a two-fold way: by their position in the hierarchy of the sciences – the more successful natural sciences produce images of nature that have lasting repercussions, for example, on the images of society – and by their complex relationship towards society and the often heroic role played by the social sciences in this domain. The present paper explores these issues and ends with the proposition that the social sciences have an important contribution to make in furthering the collective self-understanding, including that of the sciences within the realm of culture.

Any discussion of images of science is beset with the tension arising from the familiarity which shrouds the intuitive understanding of the processes through which images are generated, transmitted and received, and the attempt to render more precise what is inherently vague and transient. Images are rooted in the realm of imagination where fancy and fantasy tread a fine line with reality. They are meant to communicate something, yet the form of communication is deliberately designed to evoke responses carried by feeling and reasoning alike. One might, therefore, conclude that images are something scientists should keep away from as they yield a dangerous topic which runs the risk of leading into a realm difficult to speak about, one which is better left to literature, the autobiographies of eminent scientists, or to public relation strategists.

Yet, here we are to seriously discuss a topic that is moving us sufficiently to have come together for this occasion. Why is this so?

One of the reasons, I suppose, is that images of science are an integral and irrevocable part of the ways in which the scientific mind works. Images have their

Paper prepared for the Second Colmar Colloquium of the European Science Foundation "Image of Science – Scientific Practice and the Public". Colmar, 17–19 April 1985.

place within science. Preconceptions and non-verbal intuition are as old as science itself if we understand by them those mental constructions that operate in the 'nascent' moment in which a scientist's creative insights are shaped before they either erupt suddenly, or emerge gradually into a more conscious form. When, on February 10, 1605, Johannes Kepler revealed his devotion to the image of the universe as a physical machine in which universal terrestrial force laws were held responsible for the operation of the whole cosmos, this was – retrospectively – only one part of the imagery that moved his creative insights. His efforts would have been doomed to failure had he not supplemented the mechanistic image with two other quite different ones: the image of the universe as a mathematical harmony and that of the universe as a central theological order.[1] Similar accounts attempting to reconstruct how insights are created and what is seen before it becomes verbalized as part of the scientific language, constitute the case material on which the literature on scientific creativity draws: an hommage to vision, revealed through introspection, yet fleeting and fragmentary as an account, couched as it was into the depths of the unconscious which permits only glimpses but no full exposure of its operations.

Yet, when speaking about images we do not mean so much these pre-verbal forms of creativity, but impressions which the sciences make, especially in the mind of the public. We speak of the existence of public images of science. They function as a projection screen for the collective representations of what science is, or ought to be. These images, for reasons that have more to do with the ways in which science has been diffused differentially and accepted than with its actual development, are composed of a relatively small catalogue of stereotypes: analysis reveals that they vary between the good and the bad, between trust in science as a problem-solver and in deep anxieties concerning the uncontrollable results brought about by science and technology.[2]

Public images indicate therefore more than anything else the place assigned to science among other co-existing cultural resources and symbolic means of orientation. They reveal emotional responses on the part of the public where scientists, often misplacedly, expect a lay reflection of their own self-images.

Yet scientists, as a community organized along disciplinary lines within research fields which have their own characteristic object and mode of inquiry, not only hold images of these fields collectively, but also produce them. In these productions, science is used as a cultural resource and to variable degrees, scientific authority and expertise are brought into play. Pretending that image production is free from any vested interests would either be naive or hypocritical. Images are evoked in order to convey a message or to advance certain arguments: the image is shaped accordingly. The public appeal for research funds – to cite a frequent if somewhat trivial example – is often accompanied by images designed deliberately to evoke emotions and to stimulate the desired action.

Hence, it is difficult to separate thematic content and the ways in which images are formed, from the uses they are put to. Since use always is context-depended and since it presupposes an audience, images have to retain a certain degree of malleability. They must remain transient. For, only if they lend themselves to instant transformation into another one of their polymorphic states, can they fulfil their context-dependent, communicative function.

Images are socially produced and can only function in transmitting their intended message if they are shared. This means that they delineate for the audience with whom they are shared, also what is admitted as source of legitimate knowledge. They determine what will be considered as important, interesting, worthwhile, risky, symmetrical, beautiful, absurd or harmonious.[3] In other words, images establish a terrain of communication which includes criteria that are difficult to communicate otherwise.

The social sciences, the dominant imageries which I am expected to represent here, qualify – as stated in the beginning – both for producers of images and for producers of knowledge about images. Knowledge about the meaning and function of symbols and of symbolic means of communication, of rules of construction and de-construction, of interest-linked use and of the ritual occasions under which they become effective, largely fall within their sphere of competence. One should, therefore, expect that images produced in the social sciences – seen by the scientist – exhibit some degree of self-referential application of knowledge in this domain, and that the self-images held by social scientists, reflect their double vantage point. Yet there exist also a number of tension-introducing constraints leading to the notoriously lower level of internal consensus the social sciences exhibit, especially when compared with the natural sciences. I will, therefore, first try to elucidate the boundary conditions under which the production of images in the social sciences proceeds in order to examine later their specific vantage point.

Heroism and the Social Sciences

The constraints operate on several levels: they have evolved historically and are structured through a field of competition in which the social sciences have to contend not only for their place within the hierarchy of the sciences but also for their recognition in a socio-political space where they are confronted with widely differing expectations. The natural sciences, as Pierre Thuillier put it, operate in their imagery somewhere between God and the Devil; their representatives constantly make overt and not-so-overt use of the sacred.[4] Compared with the natural scientists and the metaphysical adornments with which they can sur-round themselves, the social scientists appear more like defrocked priests. Children of the Enlightenment, they remain bound to what they interpret to be their historical mission: to observe and interpret the ongoing project of modernization which is accompanied by a process of secularization in which God and the Church have been substituted by Society and the State.

Contrary to the natural scientists, their social counterparts are not in control of the processes they study, nor can they ever hope to be. Even if their understanding of the nature of the processes that underlie societal transformations and the evolution of social structures are relatively advanced, action − and control − are reserved for others. In addition, social scientists can equally engage in making predictions, which in itself probably is an over-estimated virtue ascribed to the natural sciences. But again, unlike the natural sciences, the predictions that have been made, can influence the outcome and thus lead to self-fulfilling, or self-destroying prophecies. Finally, the social sciences are deeply torn between the desire to be socially useful in the many areas in which social problems exist − be it poverty and unequal distribution of societal resources, inner-city riots, budgetary deficits, reading disorders in children or international conflicts, to name but a few, and their confrontation with reality in which only few solutions have a chance of being accepted and implemented. It is not only the lack of theoretical understanding of either such problem areas, nor of societal processes in general, which is at stake, nor is the usually alleged immaturity of the social sciences the crucial point. Rather, is it their position in a world of social action, and hence their reflexive and analytic task which makes it so painfully difficult to adopt a role similar to that played by the natural scientists. Being unable to act like God or the Devil or like Faust inbetween, social scientists have to invent and carefully manage the public stage on which they are to appear. Not surprisingly, if so much is left open,they also have difficulties in agreeing on the role to play, just as it is difficult to agree on what to recommend in a world in which many courses of action can be derived and justified from the analysis of the same set of facts.

Caught in the dilemma between action and reflection and hence being unable to fulfil a heroic role which comes naturally to them, social scientists have to choose between two fundamental options: they can either adopt a thoroughly unheroic posture which, at some point, becomes heroic itself, or they can opt for heroism by proxy. In either case, there is a hidden, second option lurking behind the scene. There is a choice to be made as to which social actors they wish to associate with or as to how they want to keep away from any association. In the latter case, the outcome either is the adoption of the role of the shrewd yet distant observer and commentator who maximizes credibility by refraining from taking messy action of any kind, or that of the observer who suffers from knowing about the futility and hence impossibility of intervention. In the other case, by associating themselves with the actors of the real social world, social scientists may end up on either side of the imaginary fence that runs through this world: viz. either on the side of the oppressed who cannot speak for themselves, or on the side of the alleged wielders of power and makers of decisions, as advisers to the Prince.

On the whole, economists have opted for the policy-advice-giving role. In their self-inflicted utility, as Schumpeter put it, they have become the closest approximation to the more inherently heroic figure of the natural scientist although their unanimity in matters of public policy often is nothing but wishful projection of what they perceive as theoretically and empirically more advanced knowledge base.[5] The opposite stand is taken by those social scientists who see themselves as vanguard or catalysts of social movements and who want to function as advocates for the underprivileged. They run the risk of merging their own performance with that of politics, thereby loosing their scientific identity and credibility, while the advisers to the Prince risk being subjected to the caste's periodical shake-ups and changes occurring at the Court. Nevertheless, there still remains the option that their wares might be accepted at another Court.

Not surprisingly perhaps, all of the attempts that were made by social scientists in order to enhance an inherently unheroic role by either associating themselves with the historical actors on the stage or by heightening their aloofness from them as part of a calculated strategy, have been fiercely resisted by yet another role: that of the social critic. In adopting a radical anti-heroic stand, they interpret their task as essentially one of de-mystifying the constructions with which institutions guard their privileges, by revealing the hidden functions served by rituals and by uncovering the vested interests that move thought and action. To speak up against the symbolic violence of institutions and to point out the recurrent patterns in the construction of social reality through which inequalities are maintained and created, can, however, be costly: it is at the price of painful, critical distance. "One does not enter sociology without tearing the adherences and adhesions with which one ordinarily is tied to groups; without abjuring the beliefs that constitute belonging and without renouncing all ties of affiliation . . ." was Pierre Bourdieu's "leçon sur la leçon" when he made his inaugural speech at the Collège de France. Critical distance, he maintains, should not be interpreted merely as a concession to the pervasive anti-institutional mood of the times. For Bourdieu it constitutes the only way to escape the "systematic principle of error" which lies "in the temptation of the sovereign vision". "Whenever the sociologist arrogates the right that is sometimes granted to him, namely to pronounce himself on the boundaries drawn between classes, regions and nations, to decide with the authority of science, whether social classes exist, or not, and how many, whether this or that social class – the proletariat, the peasantry, or the petite bourgeoisie . . . is reality or fiction, the sociologist usurps the function of the archaic REX . . ." Even if the threat consists in annihilating the beliefs which are the normal condition for the functioning of an institution, the illusion of knowledge exerted in the name of the "sociologist king" has to be resisted, because otherwise sociologists become guilty of complicity with the very powers that pervade society and the working of which they denounce.[6]

The dilemma is real and not only sociology – as seen by its theoreticians and practitioners – is torn between reflection and action, between engagement and distance. The field in which the social sciences operate, is itself full of conflicts that leave their trace in the way how the social space is conceptualized, in which those social actors move that are either subjects of research, sponsors, clients, or both. Policy-makers and the administrative-political establishment expect knowledge that is useful to them and policy research is located in places that can serve these expectations. Those social scientists whose ambitions have not been to advise the Prince and who have sided with the People, have raised their voice in the wake of social movements, accordingly. Although the problem may be less pronounced in disciplines that have renounced the critical distance in favour of serving the commonwealth, the balance between engagement and distance remains to be established.[7] The self-image of the social scientists is in this respect a highly unstable one: its oscillations are determined by the relations, real and imagined, that tie the social sciences to their subject of investigation: human beings who form an active part of the social reality that is analyzed.

The Construction of Order in Nature and Society

Boundary conditions for the social sciences also arise from their position within the hierarchy of the sciences. Placed midway between the highly prestigious and successful natural and life sciences and the more sheltered humanities, they are subject to often self-imposed pressures that stem from imitation and the search for a reference point of what passes as 'scientific' outside their own domain of inquiry. The well-known fact that the level of consensus within the social sciences is much lower, has many roots. One of them has been singled out by Stephen Toulmin: while the subdivision of the physical and biological sciences into largely independent subdisciplines rests on a genuinely functional differentiation between their respective problems and issues, the fragmentation of the social sciences rests – too often – on nothing more respectable than sectarian rivalry and incomprehension. Toulmin believes that in the natural sciences there has grown up by now a common and shared view of "natural philosophy", in terms of which scientists can see how their own problems are differentiated from, yet related to, those of other scientific subdisciplines. Lacking such a functional rationale for their division of labour, each group within the social sciences dreams of expanding the area of its concerns, convinced of the importance of its own problems and the value of its methodology, while questioning the significance of other approaches.[8]

Tending more towards imitating in bits and pieces what appear to be the secrets of success of the natural sciences than towards building up a coherent and internally fruitful influence between various disciplines and subdisciplines of the social sciences occurs nevertheless, especially in boundary areas. Yet, when searching for the grand images of society, fragmented and broken as they may appear when looked upon in detail, one soon discovers that such images are not free from the views and visions of nature embodied in images emanating from the natural sciences. Human philosophy, it seems, cannot exist without taking into account natural philosophy. In other words, the production of images of society does not lie in the sole sphere of competence and imagination of the social sciences. The reason for this relativized capability to produce images is rooted in the workings of the human mind. It arises from the attempts to fill the space that exists between Nature and Culture; between Humanity and the Environment; between the Natural Order and the Social Order. Images of society, just as images of nature, are constructions of order and not surprisingly, the order that has to be found in nature, continues to cast its normative shadow on the images formed of society.

Even the crudest attempt to retrace the lines of development of human thought in its effort to put order into domains that interact, but are also perceived as being distinct from each other, would amount to the impossible task of reconstructing the history of social thought and of the philosophy and history of science alike. From the earliest cosmologies onward, through the chain of material representations into which they have been cast and the marvellous remnants of which we still admire today, in all great civilizations analogies have been drawn between the social world and the world of nature. The social order has been interpreted as a representation of the Heavens and anthropomorphic representations taken from the social universe have been projected into a world moved by deities, in which the forces of nature also assumed human traits. The magic predecessor of technology were rituals invented to harness some of the cosmic energies for the purpose of social use. From simple analogies and anthropomorphic projections human imagination worked its way through even higher levels of abstraction and to the power of conceptual synthesis. Yet in all these efforts, traces remain that seek to tie together what the analytic mind is continuously separating again.

The rise of modern science is intimately linked with the search for order in nature as sharply contrasting with the unrest and social disorder that prevailed not only in 17th century England. From Kopernicus onward who put the sun on a royal throne and made it "govern the family of planets revolving around it", the universal laws of nature were extended and applied to society and the human beings living in it. Celestial harmony was translated as implying harmonization of the basic order in communities and the discovery that just one simple law creates and maintains a harmonious cosmic state, could not but leave deep repercussions on modern political thought. Newton's mechanistically unified universe, summarized later and completed by Laplace, was to exert a dominant influence on political and utopian thought well into the 19th century.[9] Even the French Revolution, which disrupted the geometrical architecture of the presumed equivalence of the physical and moral worlds, was greeted as being in accordance with the laws of nature. Condorcet and Laplace endorsed the view that force was merely taking a different course since it was now directed against those elements which impeded natural movement. Revolution became the liberation of natural harmony from feudal obstacles.[10]

So gripping is the imagery of society fashioned after the dominant imageries of nature, that the guiding analogies may change, but not the underlying predisposition. This suggests that societal arrangements should somehow conform to the patterns found within the realm of nature. The latest image of nature that exerted its dominant influence on the image of society was, of course, Darwinism. Despite the efforts towards constraining its paradigmatic effects within the realm of biology, its spill-over on society could not be avoided.

Although it is not a moot point to insist on the fact that normative prescriptions, deduced from the results of scientific research, do not fall within the domain of legitimate applications of such findings, practice continues to prove the contrary. Earlier political uses – and blatant abuses – to which Darwinism has been put, for example, in the form of Social Darwinism and its appropriation by political movements, have not prevented more recent flirtations with surrounding socio-biology, where 'laws', found in animal behaviour, are once more being transferred to the social behaviour of human beings. While most scientists coming from the physical and the biological sciences, would honestly insist on drawing the limits of their scientific expertise where the realm of political inferences begins, the line where analogies are drawn purposefully, insinuated or merely happen, is an extremely thin one in practice.

Appeals to the "natural" – and condemnations of what is considered to be "unnatural" are, as Cameron and Edge point out in their succinct study of scientism, among the most powerful forms of persuasion in the repertoire of human rhetoric. Patterns perceived in nature suggest patterns of preferred human behaviour since a natural state of affairs is assumed not merely to exist, but to exist for good reasons.[11] From her anthropological experience, Mary Douglas draws many examples for how an appeal to nature has frequently been used as a "doom point" in order to impose moral constraints.[12] Last but not least, recent feminist scholarship, especially in the field of biology, has uncovered the persistence with which a male scientific view has distorted the biological nature of women in order to fit them into the prescribed social model of behaviour and being.[13]

We have to conclude that despite their faulty logical basis which has been exposed to severe criticism on the part of philosophers, normative appeals drawn from the factual description of Nature, continue to be made. Albeit social scientists and historians of science have analyzed in detail the specific political and social conditions under which analogies between the natural and social order become fuel for political causes and conflicts, it seems extremely difficult to guard oneself against them. And it is also easy to see why: the authority of science constitutes a powerful resource to be also employed in the political arena. The more scientific expertise is embattled in questions that have become the object of fierce political controversy, the more certain images and preferred states of nature are selectively singled out. Conflicts about political decisions to be made in the field of technological development and environmental protection, to take a crucial issue of today, are not only fought with conflicting arguments and research findings in areas of scientific uncertainty, but also with conflicting images of nature, society and what their interaction should be like.

It would be a mistake, however, to reduce the problematique of the influence of images of nature on images of society solely to that of their scientistic use or abuse. Beyond the normative dimension and questions of its legitimacy or faulty logic, very powerful mechanisms operate that seek to bridge the space separating nature and society; environment and human beings. The interstice between these conceptual domains is the one in which new similarities and differences are constituted permanently and in which the search for some kind of synergistic vision continues. Whatever way the links between these domains are conceptualized, they cannot fail to have deep repercussions on scientific and technological practice alike. The power of images thus formed does not arise from its epistemological grounding, nor from any scientific reasoning alone. Rather does it stem from the emotional appeal to bring together, however tenuous and temporary the outcome of what science is separating otherwise.

The changes in the reigning images that have radiated out of science into culture from the middle of the 19th century until this day, have been extremely well described by Gerald Holton.[14] From the finite universe in time and space that featured a static, homocentric, hierarchically ordered and harmoniously arranged cosmos, rendered in delineated lines as those of Kopernicus' own hand-drawings to the "restlessness" it displayed in the last half of the 19th century, the universe is undergoing transformations that cannot fail to influence the place of science within culture. According to Holton, we are faced now with a new image, "that of a mandala feared by those critics who have never forgiven

159

science its demythologizing role – the labyrinth with the empty center where the investigator meets his own shadow only and his blackboard with his own chalk marks on it . . . his own solutions to his own puzzles."

The social sciences have long been confronted with a similar experience: while not exactly meeting their own shadow, they are constantly confronted with the problem of meaning and the shadows cast by their own constructions. Their involvement in, or detachment from them is part of the problems of the limits of their scientific expertise.

Limits of Scientific Expertise: A Classical Answer

Like any classical piece, the one I want you to get acquainted with, bears both, the date of its time and the timeless personal signature of the master. The social sciences are presented in it as an integral part of the sciences in the sense of Wissenschaft. No concessions are being made to their alleged immaturity, nor is a plea enounced for any separate treatment. The place is Germany, the time 1918, i. e. the date of Germany's transition from a disastrous war into an unstable experiment with democracy which was to end 15 years later, with the rise of fascism. The master is Max Weber speaking in front of a crowd of young people at the University of Munich.[15] The external conditions for a university career are portrayed by him as dismal: the young man – the question of women's access to the universities is not even mentioned – was in need of considerable financial resources of his own and faced an appointment structure that was described as a "mad hazard". Chance, and not talent, were determining the admittedly underserved fate of many. Mediocrities were dominating at the universities. Competition for the enrolment of students has reached ludicrous proportions putting a premium on so-called popular teachers. The responsibility of encouraging someone who is seeking advice, could hardly be borne: ". . . if he is a Jew, one naturally says 'lasciate ogni speranza'", and: "Do you in all conscience believe that you can stand seeing mediocrity after mediocrity climb behind you year after year without becoming embittered and without coming to grief?"

In stark contrast to these external conditions, Max Weber is very clear on the inner predisposition needed to enter science. Whoever is unable "to come with the idea that the fate of his soul depends on whether or not he makes the correct conjuncture at this passage of this manuscript, may well stay away from science". What was needed was "this strange intoxication, ridiculed by every outsider" . . . "for nothing is worthy of man as man unless he can pursue it with passionate devotion".

The Protestant Ethic in its incarnation as the inward calling of science, is presented as the only legitimate route of access in science. Scientific expertise in society, so the classical answer, should only be claimed by those who are unable to meet the severe preconditions of science as a vocation. Yet, such a stern but noble answer is threatened by another notion, fashionable "nowadays in circles of youth . . . the notion that science has become a problem in calculation, fabricated in laboratories or statistical filing systems, just as in a factory, a calculation involving only the cool intellect and not one's heart and soul". The idols of the cult which is under attack here, are easily named. Citing them under quotation marks suffices for everyone in the room to understand what is meant: "personality" and "personal experience". "People belabour themselves in trying to 'experience' life – for that benefits a personality, conscious of its rank and station" is Weber's acerbic comment.

The limits to scientific expertise that Weber draws – though he does not use these words – emerge from his integration of the nature of science. None of the sciences, he insists, can contain an answer to the question raised again by the youth of his days – what shall we do and how shall we live? The craving of youth for 'experience' and 'meaning' is bound to end in romantic irrationalism. The sciences presuppose as self-evident that it is worthwile to know, and while part of this knowledge is for the sake of technical results, there is also knowledge for its own sake. This presupposition, however, cannot be proved, nor can be proved that the existence of the world, as described and interpreted by science, is 'worthwhile' and that it has any 'meaning'. The genuine academic teacher – as opposed to the academic prophet and demagogue – knows this and refuses to yield to the temptation. Science can only contribute clarity – to make the necessity of choices intelligible, but it cannot make choices.

These are indeed classical issues with a long-standing tradition in philosophy and in the history of the social sciences. Re-reading Weber, one becomes aware of his lucid premonitions that were only too brutally fulfilled in the course of the subsequent political events. If Weber was right in seizing the many facets of what he called the ongoing process of rationalization and intellectualization which would lead to the thorough disenchantment of the world, he was also right in stressing that human beings are suspended in webs of significance spun by themselves. But, somehow, we have installed ourselves in a disenchanted world and while the quest for meaning continues as part of a perennial question, we are also able to invest with new meaning what irrevocably has been lost before. The criteria for recruitment into science can no longer be justified with the "inward calling" that Weber and many others would have like to be the only principle of recruitment. As I have tried to show elsewhere, it does not only need good men to do good science.[16] The organizational structure of the scientific enterprise has long since left the age of the 'noble scientist' whose motivation and stern self-discipline Weber so well portrayed himself. In an age of scientific management in which research has to be planned well in advance, and in which the enormous sums invested in a single experiment necessitate an equally enormous effort in managerial skills and in mastering organizational complexity, the inner devotion alone appears hopelessly obsolete. If it is still a motivating force, the inner passion has to be organized collectively.

Today, the problem of the limits of scientific expertise presents itself in very concrete and new general terms. On the level of professional competence and competition, the limits that have to be re-drawn, mainly arise from increasing interaction with a public that no longer seeks 'meaning' from science. It also demands its share in the decision-making process with regard to scientific and technological developments or asks for greater accountability in the interaction between laymen and -women and the experts. The 'meaning' which is demanded here, is broken down into a myriad of concrete and conflicting components: as seen by the decision-makers who operate in different policy fields, or as seen by the public; as seen in terms of maintaining economic competitiveness or the ecological balance; or as judged through the relation of ends and means. Each of these criteria can be questioned as part of the policy process. In the daily practice of decision-making and of the impact of scientific expertise preceding or following it, science for public policy necessarily is subject to constant negotiations as essential part of the political process which generated scientific expertise in the first place.

I cannot see any inherent difference here between the social sciences and the rest although the areas of competence and the institutional locations are bound to differ. The growing incorporation of science into the political and especially into the military field of application equally forces scientists to take a stand and to re-define the weight and bearing of their expertise under concrete circumstances. Whether this entails taking a stand with regard to the war in Vietnam or the war of the stars, whether scientists find themselves entangled as experts pro and contra in public controversies about nuclear power, in decision-making processes bearing on environmental protection or on occupational hazards arising from the manufacture of chemicals, the line between facts and values – whose protagonist in the social sciences Max Weber was – has turned out to be a very thin one, also in the other sciences. Scientific expertise has become a political resource and its limits are thus subject to the limits or boundaries that separate science from politics.

Experience, We and the Other

Elaborating the constraining influences under which the social sciences operate in their production of images and which – in this respect at least – differentiate the social sciences from other sciences and simultaneously insisting on very similar conditions that determine, e. g. the limits of scientific expertise, I have come to the last point: what are some of the continuing social functions of images of society and where do images of society and images of nature intersect and mutually feed upon each other?

Such questions, inevitably take us into the realm of culture in which science figures prominently as one form of symbolic expression. The official relationship between culture and science has not been an easy one, especially when culture is treated in the narrower sense of representing the arts only. The modernist dilemma, as David Dickson puts it, dates back at least to the middle of the past century when the modernistic spirit was born out of cultural struggles and in particular out of an effort towards defining the arts as a sphere of consciousness that stood in opposition to the social impact of industrialization and technological change. Art was seen by some as an appropriate and necessary way of adjusting the growing imbalance between material and spiritual (or even moral) progress – a luxury item to be enjoyed in moments of leisure and relaxation by those who saw themselves as the vanguard of progress to be achieved through scientific and technological means. It was seen as a fighting ground for others who were convinced that they had to preserve humanistic wisdom and values while exposing themselves without reserve to a form of existence which sought to render conscious as an equally legitimate form of expression what otherwise would simply be repressed and referred to marginality. The metaphysical split, so well described by A. N. Whitehead, between the objective characteristics of the world out there accessible to all, quantifiable and inter-subjectively verifiable and the subjective experience of the 'inner world' of emotions, feeling and subjectivity, unfortunately was even further deepened by association with the domains of the rational and the irrational. The central dilemma, according to Dickson, arises out of the fact that the modernist movement has sought not to deny this split, but to exploit it.

It is my conviction that the social sciences have a potentially important contribution to make in analyzing and de-mystifying this split, and in eventually helping to overcome it. Max Weber's stern answer to those young people in his audience who were 'craving for experience', suited as it was to the occasion, has long since given way to the study of subjective experience as a central object of analysis in the social sciences. Experience has been incorporated, in its individualistic expressions as well as in its collective ones, by almost all disciplines. Historians, for instance, have become deeply intrigued by describing and reanalyzing the experience of ordinary men and women as a crucial part of the historical process and several of the sparse documentary accounts that exist, probing into their perception and inner world of representation, have turned out to be best-sellers. Oral history has likewise become a complementary method by now indispensable in recording living historical witnesses' accounts for the enlightenment of future generations. The work of anthropologists dealing with the rich symbolism of rituals, to take yet another example, would be reduced to the mere analysis of functions served by them, were it not for incorporating the meaning that practitioners attribute to such rituals. Everyday life experience has likewise become an almost fashionable topic in some quarters of the social sciences while even economics has to deal with certain aspects of subjective judgement, preferences and utility functions in areas like investment and consumption behaviour or how individuals cope with their uncertainty.

The study of experience is, of course, not to be equated with experience as such. Nor can the study of experience by-pass the process of abstraction and the tendency towards generalization inherent in the scientific method. Nevertheless, the gradual inclusion of experience as object of study has slowly led towards a transformation of views about the construction of social reality, and of how the 'inside' conditions are linked and interwoven with those that are assumed to be 'outside'. The split between the subjective and objective side undoubtedly has become blurred. Gradually, the individual side of otherwise collective phenomena becomes understandable in such terms, just as the aggregation and interlinkage of individual actions into collective patterns of behaviour become more understandable in their interrelationship. In this sense, one can say that it is the task of the social sciences to expand our collective self-understanding by being able to make the experience of the I intelligible and to translate it into the experience of the We and of the Other. Even if most processes in the social world are blind ones inasmuch as they are directed – their final outcome for the most part, however, being unintended by anyone as such – we continue, through individual beliefs, emotions and actions, to be a shaping component of these processes. Even if we hold our highly personal experience to be unique, it nevertheless reveals itself as being part of a larger pattern in which we behave qua social beings. Collective self-understanding is that process through which the transformation mechanisms, separating the unique experience of the I and the shared experience of the We and the Other, become understandable in terms of both.

Images of society refer to collectively shared experience and it is in this quality that social experience enters, albeit surreptitiously, images of nature as well. The central vehicle through which the social world impinges upon the realm of nature is thought and language. Concepts, developed with the aim of explaining nature, nevertheless derive part of their meaning from social contexts and rely on shared understanding. Social arrangements have left their traces historically in the development of scientific concepts – a process which continues to this very day, however abstract and context-independent scientific concepts eventually may become. Recent work in ecology, for example, favours concepts such as 'resilience' of natural systems, their propensity towards 'surprise' or their 'vulnerability' in charting the still largely unknown terrain between environment and human forms of intervention.

In elucidating the symbolic processes through which such transfers are operating and analyzing the roots of experience that scientists – all scientists – continue to have in the social world, the social sciences can contribute to the joint endeavour of collective self-understanding. The areas of overlapping imageries, of conflicting claims and of shifting dominance are in this context far more revealing than the content of such images when treated separately. It is the space in-between, and especially in-between nature, society and the biological condition that remains to be explored.

Furthering collective self-understanding for the social sciences does not preclude falling back into attempts to turn an essentially unheroic role into a heroic one. But independently of the specific style in which they proceed, if joint progress could be achieved in this area, the images of science will turn out be an excellent guide for finding and redefining anew the place of science – of all sciences – within culture today.

References

1   Gerald Holton. Introduction. in: Thematic Origins of Scientific Thought. Cambridge Mass.: Harvard University Press. 1973.
2   Gerald Holton. 'Modern Science and the Intellectual Tradition'. ibid.
3   Yehuda Elkana. 'A Programmatic Attempt at an Anthropology of Knowledge'. in: Everett Mendelsohn and Yehuda Elkana (eds.) Sciences and Cultures. Sociology of the Sciences. vol. V. 1981 Dordrecht: D. Reidel Publishing Company.
4   Pierre Thuillier. Les savoirs ventriloques ou comment la culture parle à travers la science. Paris: Seuil. 1983.
5   E. Malinvaud. "The Identification of Scientific Advances in Economics". in: Torsten Hägerstrand (ed.) The Identification of Progress in Learning. Cambridge: Cambridge University Press. 1985.
6   Pierre Bourdieu. Leçon Inaugurale. Le Monde 25 – 26 avril 1982.
7   Norbert Elias. Engagement und Distanzierung. Frankfurt a. M.: Suhrkamp 1983.
8   Stephen Toulmin. "Towards Reintegration: An Agenda for Psychology's Second Century". in: Richard A. Kasschau and Charles N. Cofer (eds.) Psychology's Second Century: Enduring Issues. New York: Praeger. 1981.
9   Michael Winter. "The Explosion of the Circle: Science and Negative Utopia". in: Everett Mendelsohn and Helga Nowotny (eds.) Nineteen Eighty-Four: Science between Utopia and Dystopia. Sociology of the Sciences. vol. VIII. 1984. 73 – 90. Dordrecht: Reidel Publishing Company.
10 Michael Winter. ibid.
11 Iain Cameron and David Edge. Scientific Images and their Social Uses, An Introduction to the Concept of Scientism. SISCON. 1979.
12 Mary Douglas. "Environment at risk". in: Implicit Meanings, London: Routledge and Kegan Paul. 1975.
13 See. for instance, Bleier Ruth. Science and Gender. London: Pergamon Press. Athene Seria. 1984: Janet Sayers. Biological Politics. London: Tavistock. 1982.

14   Gerald Holton, op. cit. note 1.

15   Max Weber, "Science as a Vocation", in H. H. Gerth and C. W. Mills (eds.) From Max Weber: Essays in Sociology. Oxford University Press, 1985 ('Wissenschaft als Beruf', Gesammelte Aufsätze zur Wissenschaftslehre, Tübingen, 1922).

16   Helga Nowotny, "Does It Only Need Good Men to Do Good Science?", in Gibbons M. and Wittrock B. (Eds.), Science as a Commodity, Longman 1985.

STRANGER THINGS

INTO THE FIRE

BOYD COUNTY

script
JODY HOUSER

pencils
RYAN KELLY

inks
LE BEAU UNDERWOOD

colors
TRIONA FARRELL

lettering
NATE PIEKOS OF BLAMBOT®

front cover art by
KYLE LAMBERT

chapter break art by
VIKTOR KALVACHEV

dark horse books

president and publisher
MIKE RICHARDSON

editor
SPENCER CUSHING

assistant editor
KONNER KNUDSEN

collection designer
PATRICK SATTERFIELD

digital art technician
ALLYSON HALLER

Special thanks to NETFLIX, SHANNON SCHRAM, ANASTACIA FERRY, and KYLE LAMBERT.

Advertising Sales: (503) 905-2315 | ComicShopLocator.com

This volume collects issues #1 through #4 of the Dark Horse comic-book series
Stranger Things: Into the Fire.

Published by Dark Horse Books
A division of Dark Horse Comics LLC
10956 SE Main Street
Milwaukie, OR 97222

DarkHorse.com | Netflix.com

First edition: August 2020

Ebook ISBN: 978-1-50671-317-5
Trade Paperback ISBN: 978-1-50671-308-3

1 3 5 7 9 10 8 6 4 2
Printed in Canada

"I THINK IT'S *PERFECT*, MARCY."

I DON'T KNOW...

HERE, LET'S TRY THIS.

I KNOW YOU HAVEN'T BEEN AT SCHOOL *THAT* LONG.

BUT YOU *SHOULD* KNOW BETTER THAN TO QUESTION SARAH'S TASTE BY NOW.

WHY DO YOU THINK DAWN LOOKS SO GOOD? SHE WAS *HOPELESS* BEFORE WE WERE BEST FRIENDS.

I WASN'T *THAT* BAD, THANK YOU.

NO, YOU'RE RIGHT. IT'S PRETTY GREAT.

AND YOU *REALLY* NEED TO UP YOUR FASHION GAME IF YOU WANT JOSH TO NOTICE YOU.

I DON'T... WHAT?

YOU MIGHT BE A *LITTLE* LESS OBVIOUS IF YOU WEREN'T STARING AT HIM LIKE *ALL* THE TIME.

THE ONLY REASON HE HASN'T NOTICED IS HE'S *KIND* OF AN IDIOT.

HE'S NOT AN IDIOT.

I SAID "KIND OF."

WE CAN INTRODUCE YOU, IF YOU INTRODUCE US TO YOUR BROTHER.

UGH, DON'T BE GROSS, YOU GUYS. HE'S LIKE *TWENTY-FIVE.*

FILENE'S

12

"I SPENT YEARS THINKING THE PEOPLE WHO RAN THE PROGRAM REALLY WANTED TO HELP US ALL.

"BUT THE WAY THEY PUSHED YOUR SISTER UNTIL SHE BROKE...IT WAS CLEAR THEY NEVER CARED ABOUT US AT ALL.

"IF THEY COULDN'T USE US, CONTROL US, THEY'D PREFER WE WERE DEAD."

WE WERE BOTH LUCKY TO ESCAPE WHEN WE DID.

SIX... FRANCINE, SHE *DIED* TO MAKE SURE IT HAPPENED.

AND IF THERE'S *ANY* CHANCE THAT ANY OF THE OTHER SUBJECTS EVENTUALLY GOT OUT...

WE NEED TO FIND THEM. HELP THEM IF WE CAN. WE'LL BE SAFER TOGETHER.

BUT WHEN WILL ALL THAT BE OVER?

WHEN WILL WE JUST GET TO BE, YOU KNOW. *NORMAL.*

WHEN WE FIND THE OTHERS.

OR...AT LEAST FIND OUT WHAT *HAPPENED* TO THEM.

AND IF WE NEVER DO?

BETTER THAT THEN NOT TRYING AT ALL, RIGHT?

IF THEY'RE OUT THERE WAITING FOR SOMEONE TO HELP THEM...

I KNOW I'M NOT GREAT AT ANY OF THIS.

BUT WE'RE THE CLOSEST THING TO FAMILY THE OTHER HAS.

YOU'RE NOT THE *WORST* FAKE BIG BROTHER EVER.

PROBABLY.

WHA...

WHERE ARE WE?

I'M AWAKE! I'M AWAKE!

HASTINGS GLEN. WE DROVE ALL NIGHT.

WELL. I DID.

I'M ALMOST SIXTEEN. I COULD HELP WITH THE DRIVING.

IT'S NOT LIKE WE DON'T LIE ABOUT ALL THE OTHER STUFF.

MAYBE LATER.

I'M GOING TO GO SEE WHAT INFO I CAN GET FROM THE LOCAL COPS.

GO JEDI MIND TRICK THE CRAP OUT OF THEM.

IT SAYS THE PERPETRATORS WERE WEARING HALLOWEEN MASKS?

SO NO ONE KNOWS IF THEY WERE ADULTS OR KIDS?

YOU REALLY THINK *KIDS* COULD JUST PLUG A GUY LIKE THAT?

IF THEY'RE PUSHED HARD ENOUGH? SURE.

HEY, LOIS, ANY CALLS FROM--

THEY HAD ME BE A PART OF HER TRIALS A FEW TIMES.

SHE WAS TRYING TO MAKE IT SEEM LIKE SHE WASN'T THERE.

JUST THE KIND OF THING THOSE CREEPS WOULD DO TO A KID.

PROBABLY WANTED HER TO BE SOME KIND OF ASSASSIN OR SOMETHING.

LIKE SHE'S DOING NOW.

WE DON'T KNOW WHAT HAPPENED AFTER WE LEFT. HOW BAD THINGS GOT.

HE MIGHT HAVE DESERVED IT.

CHICAGO SIX MONTHS AGO. AND NOW THIS.

IT SEEMS LIKE SHE'S KEEPING A LOWER PROFILE. STAYING ON THE MOVE.

SO HOW DO WE FIND HER?

WE ASK AROUND, OBVIOUSLY.

THAT'S THE ONE I SAW IN THE SECURITY FOOTAGE.

SHK-SHK

FREEZE, CREEPS.

THIS IS PRIVATE PROPERTY. *SCRAM.*

UNLESS YOU WANT TO BE FERTILIZER.

WE'RE JUST LOOKING FOR A FRIEND.

DO YOU THINK YOU COULD HELP US?

PRETTY PLEASE?

...NOT SURPRISED WORD GOT OUT ABOUT CHICAGO. IT WAS REAL--

WHAT THE *HELL*, AXEL?

THEY'RE JUST LOOKING FOR THEIR FRIEND.

EIGHT?

JUST... JUST STAY CALM...

I **KNOW** THAT. I THOUGHT YOU **WERE** HER.

I SAW **HER** A LOT MORE RECENTLY THAN YOU, AFTER ALL.

JUST HOW MANY OF THESE KIDS **ARE** THERE?

NOT A CLUE.

BUT THE ACCIDENT...SHE WAS DYING...

SIX. FRANCINE. SHE TOLD US NINE WASN'T GOING TO MAKE IT.

I DON'T KNOW WHAT TO TELL YOU...

...BUT THE LAST TIME I SAW YOUR SISTER, SHE WAS VERY MUCH ALIVE.

"YOU KNOW PART OF THE STORY.

"THE PROGRAM THAT **PRETENDED** IT WAS HELPING US.

"THE DAY THAT WOMAN BROKE IN.

"TURNS OUT SHE WAS JANE-- ELEVEN'S MOTHER."

JANE!

"HOW A FEW OF YOU TRIED TO ESCAPE IN THE CHAOS.

"HOW YOU DIDN'T ALL MAKE IT.

"DID YOU EVEN THINK ABOUT THE ONES YOU LEFT BEHIND?"

33

I WANT MY FRIENDS!

JESUS...

DID THEY...

...DID ANYONE SAY WHAT HAPPENED TO HER?

NO. AND I CAN'T SAY I REALLY CARED AT THE TIME.

SHE SCARED ME.

BUT BETWEEN THE ACCIDENT, BEING TOLD SHE WAS ABANDONED, THE ILLUSIONS, WHATEVER DRUGS THEY HAD HER ON...

...IT'S NO WONDER SHE SEEMED CRAZY.

THAT PLACE BROKE KIDS, YOU SHOULD HAVE SEEN ELEVEN.

SHE COULD BARELY--

WAIT, ELEVEN? DO YOU KNOW WHAT HAPPENED TO HER?

SAW HER A FEW MONTHS BACK. SHE GOT OUT. TRACKED ME DOWN WITH HER POWERS.

"DIDN'T BOTHER TO STICK AROUND. HAS FRIENDS SOMEWHERE."

SEEMED LIKE SHE WAS DOING JUST FINE WITHOUT ANY OF US.

BUT FROM WHAT YOU SAID, JAMIE *WASN'T.*

DO YOU HAVE *ANY* IDEA WHAT COULD HAVE HAPPENED TO HER? *ANYTHING* THAT COULD HELP US FIND HER?

I MAY HAVE SOMETHING...

WHERE DID YOU GET THESE?

NONE OF YOUR BUSINESS.

A REPORTER NEVER REVEALS HER SOURCES!

WE AREN'T EXACTLY REPORTERS, DOTTIE.

HERE. THIS IS THE DOCTOR WHO WAS WORKING WITH NINE THE MOST, LAST TIME I SAW HER.

IF ANYONE KNOWS WHAT HAPPENED TO HER, IT WOULD PROBABLY BE HIM.

GOOD THING WE HADN'T GOTTEN TO THAT ONE, YET.

QUIET, AXEL.

"GOTTEN TO?"

THE PROGRAM HURT US. MY FRIENDS AND I HURT THEM RIGHT BACK.

ONLY WE MAKE SURE *THEY* CAN'T HURT ANYONE EVER AGAIN.

YOU'RE WELCOME TO JOIN US AFTER YOU FINISH LOOKING FOR YOUR SISTER.

NO... NO THANK YOU.

THANK YOU FOR THE NAME.

OPEN INVITATION.

I HAVE A NAME FOR YOU TO LOOK UP.

DR. EDWARD J. MORRIS...

JAMIE...

1975.

JAMIE AND MARCY, IS IT?

I'M SO SORRY TO HEAR ABOUT YOUR FAMILY.

YOU *CAN'T* TAKE JAMIE AWAY!

AND WHY WOULD I BE TAKING JAMIE AWAY?

BECAUSE I MADE THE FIRE HAPPEN.

AND NOW THEY'RE ALL DEAD.

AND THAT'S NOT YOUR FAULT. WE HAD APPROACHED YOUR PARENTS ABOUT GETTING YOU HELP FOR YOUR...CONDITION.

IF THEY HADN'T IGNORED OUR WARNINGS, THIS WOULD NEVER HAVE HAPPENED.

MY NAME IS DR. BRENNER. I RUN A PROGRAM FOR SPECIAL CHILDREN JUST LIKE YOU.

WE'LL BE ABLE TO HELP YOU SO THAT NOTHING LIKE THIS EVER HAPPENS AGAIN.

BUT WHAT ABOUT MARCY?

SHE ISN'T LIKE ME. SHE'S NORMAL. CAN SHE STILL COME TOO?

OF COURSE, JAMIE.

THE PRINCESS WONDERED IF SHE WOULD EVER BE WARM AGAIN.

OF COURSE, THERE WAS ALWAYS THE OLD WAY.

THE *DANGEROUS* WAY.

BUT THAT MAGIC HAD A STEEP PRICE.

SHE'S COMPLAINING SHE'S COLD AGAIN. ASKED TO SEE YOU.

WELL, LET'S TAKE A LOOK...

AND SHE WASN'T SURE SHE WAS READY TO PAY IT AGAIN. NOT JUST YET.

UNDER THE TONGUE. THAT'S A GOOD GIRL.

OF COURSE, THE PRICE COULD ALWAYS BE PAID BY THOSE AROUND HER, THE PRINCESS KNEW.

EVERYTHING LOOKS NORMAL. I'LL SEE IF WE CAN RAISE THE TEMPERATURE IN HERE A BIT FOR YOU.

AND DESPITE THE KINDNESS OF HER FEATHERED AND FURRY FRIENDS...

YOU GET SOME REST NOW.

...THE DEAL WITH THAT PARTICULAR DEVIL WAS VERY TEMPTING INDEED.

C...COLD...

GOT IT. OKAY, THANKS A MILLION.

GOOD, YOU'RE UP.

I GOT AN ADDRESS AND DIRECTIONS TO OUR DR. MORRIS.

IF WE LEAVE SOON, WE SHOULD BE ABLE TO GET THERE BEFORE TONIGHT.

G... GOOD.

YOU OKAY, MARCY?

YEAH.

JUST A WEIRD DREAM, IS ALL.

YOU OKAY? READY TO DO THIS?

I... I THINK SO.

HEY, I'M HERE FOR YOU ALL THE WAY, OKAY?

YOU DON'T HAVE TO DO THIS ALONE.

I KNOW.

ding-
dong

LIGHT'S ON.
SOMEONE'S
DEFINITELY
HOME.

HE
MIGHT NOT
WANT TO OPEN
UP WHEN HE
SEES--

GET
INSIDE.
NOW.

COME TO
KILL ME?

WASN'T GOING TO. PROMISE.

BUT IF YOU KILL HIM, WE MAY NEVER FIND YOUR SISTER.

I DIDN'T SAY I WAS GOING TO *KILL* HIM.

JUST *SHOOT* HIM.

YOU DON'T...YOU DON'T HAVE TO SHOOT ME.

I'LL TELL YOU EVERYTHING I KNOW.

WILL YOU GIVE ME THE GUN, PLEASE?

"YOUR TWIN, NINE, WAS ONE OF THE MORE PROMISING CANDIDATES IN THE PROGRAM."

AT LEAST IN TERMS OF SHEER POWER POTENTIAL.

BUT SHE HAD THE SAME PROBLEM MOST OF THE CHILDREN HAD.

"CONTROL. OR LACK THEREOF.

"AND WHEN YOU'RE USING CHILDREN WITH POWERS TO CONTROL OTHER CHILDREN WITH POWERS...

"...WELL, YOU DON'T NEED TO BE A SCIENTIST TO FIGURE OUT THAT THAT'S AN UNTENABLE SCENARIO."

JUST AS NINE WAS PROVING HERSELF TO BE QUITE A PROBLEM, DR. BRENNER HAD A BREAKTHROUGH ELSEWHERE.

ONE THAT CONVINCED HIM THE OTHER PARTICIPANTS IN THE PROGRAM WERE... REDUNDANT.

WHAT DO YOU MEAN, REDUNDANT?

THE PRICK OF A SPINDLE. THE BITE OF AN APPLE.

ANYTHING TO TAKE AWAY THE DARKNESS OF THE WAKING WORLD.

TO RETURN TO THE BRIGHTNESS OF HER DREAMS. THE WARMTH OF HER FRIENDS.

BUT NO MATTER HOW TIGHTLY SHE SQUEEZED HER EYES SHUT, THERE WAS ONLY MORE DARKNESS TO BE SEEN.

AND OF COURSE, THE DARKNESS IS WHERE THE MONSTERS LURK.

MONSTERS MORE POWERFUL THAN ANY SPELL A PRINCESS MIGHT WEAVE.

NO!

BUT WHILE HER FRIENDS WERE LOST TO THE LIGHT, THE DARKNESS HELD MORE THAN JUST MONSTERS.

THERE WERE ALLIES TOO, ALSO LOOKING TO ESCAPE TO FAIRER LANDS.

"WHEN BRENNER ASKED ME TO TAKE CARE OF THE CHILDREN HE DEEMED 'LIABILITIES', IT WAS CLEAR WHAT HE MEANT.

"I HAVE NO ILLUSIONS ABOUT BEING A GOOD PERSON. I'M SURE MANY WOULD CALL ME A MONSTER.

"BUT EVEN MONSTERS HAVE LINES THEY WON'T CROSS.

"I WORKED PRIMARILY WITH NINE, YOUR SISTER, BY THAT POINT. BRENNER WANTED ME TO...

"...DISPOSE OF HER.

"HOWEVER, I HAD A WAY TO SAVE HER LIFE, AT THE VERY LEAST.

"A DOCTOR WHO OWED ME A FAVOR WAS ABLE TO PLACE HER IN THEIR FACILITY.

"COMPLETELY OFF THE BOOKS.

"THEY AGREED TO KEEP HER ON A STRICT DRUG REGIMEN, ONE I DESIGNED TO SUPPRESS HER ABILITIES.

"AND MORE IMPORTANTLY, THEY WERE WILLING TO DO SO WITHOUT ASKING ANY QUESTIONS."

SO LET ME GET THIS STRAIGHT...

...YOU JUST DUMPED HER ON SOMEONE ELSE? DECIDED SHE WASN'T YOUR PROBLEM ANYMORE?

I *SAVED* HER! IF IT WASN'T FOR ME, SHE'D BE *DEAD* RIGHT NOW!

AND THEN YOU WOULD HAVE BEEN TOO.

MARCY...

WHAT'S THE NAME OF THE FACILITY WHERE YOU LEFT HER?

I CAN GIVE YOU THE NAME AND ADDRESS IF YOU LIKE.

BUT IF YOU'RE THINKING OF BREAKING HER OUT, YOU'D BE MAKING A MISTAKE.

YOU CAN'T GO OFF SCRIPT LIKE THAT, MARCY.

IT'S TOO DANGEROUS. YOU MIGHT--

GET HURT? LIKE THAT HASN'T BEEN MY WHOLE SHITTY LIFE?

AND WHAT SCRIPT? IT'S NOT LIKE YOU KNOW WHAT YOU'RE DOING HERE.

I KNOW. I *KNOW.*

BUT I'M TRYING. OKAY?

I'M-- LOOK--

THIS ALL JUST SUCKS, OKAY? *SO* MUCH.

BUT I KNOW YOU'RE TRYING TO MAKE IT SUCK LESS. AND SOMETIMES YOU DO.

I'M DOING THE BEST I CAN.

AND I KNOW THAT'S NOT ALWAYS ENOUGH.

BUT IF WE WORK TOGETHER, IT'S MORE LIKELY THAT IT WILL BE.

MIRRORS COULD ALSO LIE. SHE KNEW THAT TOO WELL.

BUT THIS TIME, SHE SOMEHOW KNEW THAT THE REFLECTION WASN'T THE LIE.

PERHAPS THE REFLECTION WAS THE LIAR.

BANG

THERE WAS AN OLD ANGER HERE. A BETRAYAL HALF-REMEMBERED.

BANG

THE FACE, SO MUCH LIKE HER OWN, WASN'T HERS AT ALL.

THIS FACE IN THE MIRROR HAD WOUNDED HER, LONG AGO AND FAR AWAY. A DEEP AND VICIOUS BLOW.

BANG

HOW COULD SHE HAVE FORGOTTEN?

THAT'S WHAT MAGIC DID. MAGIC AND TIME.

KRAK

BUT THE PRINCESS HAD FINALLY REMEMBERED. REMEMBERED HOW MUCH IT HURT.

REMEMBERED HOW MUCH PAIN THERE WAS TO PAY BACK.

BUT MUCH LIKE PRINCESSES, MAGIC WAS HARD TO TRAP BEHIND A DOOR ONCE IT HAD BEEN SET FREE.

WAA-BWAA-BWAA

HEY! GET AWAY FROM THERE!

67

"I'M ALL SHE HAS."

SHE HAS ME TOO. YOU BOTH DO.

I JUST WORRY THAT WON'T BE--

WEEEOOO-
WEEEOOO-
WEEEOOO

WEEEOOO-
WEEEOOO

WEEEOOO-WEE

THE PRINCESS FLED THROUGH THE WOODS, AS PRINCESSES OFTEN DO.

AND FIRE FOLLOWED IN HER FOOTSTEPS.

FOR SO LONG, FIRE HAD BEEN HER ONLY FRIEND. HER MAGIC. HER WEAPON.

HER TEMPTATION.

AND NOW THAT SHE HAD SET IT FREE, THE PRINCESS WASN'T SURE SHE IF IT WAS A MAGIC SHE COULD CONTROL...

"SEVERAL OF THE PATIENTS ARE STILL MISSING."

SHE'S NOT THERE, MARCY.

BUT THEY DON'T THINK SHE WAS INSIDE WHEN... YOU KNOW...

WHEN THE BUILDING WENT UP.

YEAH. THAT.

IF SHE RAN, SHE COULDN'T HAVE GOTTEN FAR. BUT THIS IS A BIG AREA TO SEARCH.

THE WOODS.

"REMEMBER WHAT EIGHT TOLD US ABOUT THE ILLUSIONS THEY HAD HER DO? TO KEEP JAMIE CALM?

"THE WOODS PROBABLY FEEL SAFE TO HER."

JAMIE?!

THE PRINCESS COULDN'T RUN FROM THE FIRE. SHE KNEW THAT NOW.

HER WILL. HER CURSE. HER WEAPON.

SO WHY RUN AT ALL?

FOR SO LONG, SHE HAD BEEN SCARED OF THE POWER INSIDE HER.

BUT IF SHE COULD TRULY WIELD IT...

...SHE WOULD NEVER NEED TO RUN AGAIN.

THEY HAD STRIPPED HER OF HER WEAPON.

OR SO THEY HAD THOUGHT.

BUT THE SWORD WAS JUST THE WEAPON THE WORLD UNDERSTOOD.

DON'T. YOU HAVE TO LET ME TRY.

BUT--

IT'S THE ONLY WAY TO MAKE THIS WORK.

JAMIE, PLEASE LISTEN.

I KNOW YOU'VE BEEN HURT. AND I KNOW SOME OF THAT IS MY FAULT.

THEY TOLD ME THAT YOU WEREN'T GOING TO MAKE IT.

BUT I SHOULDN'T HAVE BELIEVED IT. I SHOULD HAVE KNOWN THAT YOU WERE STRONGER THAN THAT.

I SHOULD HAVE KNOWN YOU WERE STILL HERE. WE SHOULD HAVE COME LOOKING FOR YOU SOONER.

BUT I'M HERE NOW, OKAY? I CAME HERE FOR *YOU*.

AND I'M HERE FOR YOU NOW. I'M NOT LEAVING AGAIN.

BUT YOU HAVE TO LET ME IN, JAMIE.

YOU HAVE TO HEAR ME.

THE PRINCESS HAD SHATTERED THE MIRROR ONCE BEFORE.

JAMIE...

...IT'S OKAY. YOU'RE OKAY.

I REMEMBERED YOU.

LOOK, I HATE TO INTERRUPT A TOUCHING MOMENT.

BUT WE HAVE AN ESCAPED PATIENT WHO SET A HOSPITAL ON FIRE.

WE REALLY NEED TO GET OUT OF HERE.

...THREE?

IT'S RICKY.

BUT WE CAN DISCUSS ALL THAT LATER.

"WE HAVE A *LOT* OF CATCHING UP TO DO..."

MOTEL

I'M NOT SURE IF "ALL RIGHT" IS THE PHRASE I'D USE.

BUT WE GOT TO HER. SHE'S SAFE NOW.

THEN SHE HAS A CHANCE.

THAT'S ALL I EVER WANTED FOR HER. FOR ANY OF YOU.

A CHANCE FOR *WHAT,* EXACTLY?

BECAUSE I'M THINKING A NORMAL LIFE WAS NEVER REALLY IN THE CARDS.

A CHANCE TO...TO BE *HAPPY.*

OUTSIDE THE CONFINES OF A LABORATORY EXPERIMENT.

THE LAB YOU WORKED AT? THE LAB WHERE YOU WERE ONE OF THE MONSTERS?

THE LAB YOU DIDN'T DO A DAMN THING TO SHUT DOWN?

90

Art by CHUN LO

HOWEVER, THERE ARE A FEW WHO KNOW THE TRUTH OF WHAT HAPPENED.

WHAT? PRETZELS ARE GREAT! YOU DON'T LIKE PRETZELS?

JUST... DON'T GIVE HER A REASON TO THROW YOU OUT, OKAY?

THE MONSTERS WHO LURKED IN THEIR MIDST. IN THE DARK.

AND THOUGH THEY MIGHT PRETEND...

...CAN THINGS EVER TRULY BE NORMAL FOR THEM AGAIN?

OH COME ON, YOUR MOM *LOVES* ME.

SHE *TOLERATES* YOU, AT LEAST.

I DON'T THINK SHE'S *THRILLED* WE'RE DATING.

BUT AT LEAST IT'S SOMETHING NORMAL. SOMETHING SHE *UNDERSTANDS.*

PLUS, I'M NOT LIVING IN YOUR BASEMENT.

NOT FUNNY.

MIKE IS STILL...

...HE'S REALLY UPSET ABOUT ELEVEN DISAPPEARING.

OH.

WELL, UH...WHAT DID YOU WANT TO KNOW?

THE PAPERS TALKED ABOUT WHEN HE CAME BACK FROM "THE DEAD."

BUT NOTHING AFTER THAT.

"AND WE HAVEN'T SEEN HIM SINCE THE NIGHT THEY PULLED HIM OUT OF...WHATEVER THAT WAS.

"DID HE GET OUT OF THE HOSPITAL?"

COULDN'T THEY JUST ASK JONATHAN ABOUT THIS STUFF?

YEAH. RIGHT BEFORE THANKSGIVING.

"WE WENT AND SAW HIM EVERY DAY IN THE HOSPITAL AFTER SCHOOL.

"BUT WE HAVEN'T BEEN BY HIS HOUSE YET. MRS. BYERS SAID HE NEEDED SOME TIME."

YEAH, IT WAS KIND OF...

WE MAY HAVE SET THE MONSTER ON FIRE IN THEIR HALLWAY.

WE SHOULD PROBABLY ASK IF THEY NEED ANY HELP FIXING THE PLACE UP...

LOOK, IT'S COOL THAT YOU GUYS ARE WORRIED ABOUT WILL...

...BUT IT SOUNDED LIKE THE DOCTORS THOUGHT HE'D BE OKAY.

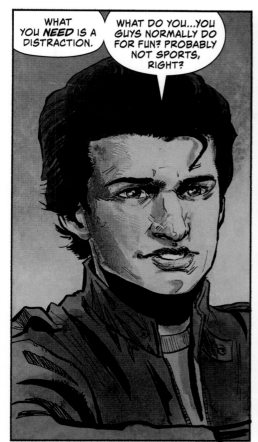

WHAT YOU **NEED** IS A DISTRACTION.

WHAT DO YOU...YOU GUYS NORMALLY DO FOR FUN? PROBABLY NOT SPORTS, RIGHT?

WHY DON'T YOU RUN ONE OF YOUR GAMES?

THE LATEST ADVENTURE IS USUALLY ALL YOU AND YOUR FRIENDS CAN TALK ABOUT FOR LIKE A **WEEK** AFTER.

WHAT?

MAYBE...

SO... THEY **DO** PLAY SPORTS?

NO, IT'S A PEN AND PAPER THING. DUNGEONS AND DRAGONS.

IT'S THIS AMAZING GAME! IT'S LIKE...

HAVE YOU EVER READ LORD OF THE RINGS?

NOPE.

SO DUNGEONS AND DRAGONS IS THIS REALLY COOL FANTASY GAME.

YOU HAVE CLASSES LIKE FIGHTERS AND MAGIC-USERS. AND YOU CAN BE HUMANS OR DEMI-HUMANS.

THE PLAYERS ARE AN ADVENTURING PARTY. FIGHTING MONSTERS AND LOOKING FOR TREASURE AND STUFF.

THE LAST TIME WE PLAYED...

...WAS THE NIGHT WILL DISAPPEARED.

IT'S NOT YOUR FAULT, MIKE.

YOU GUYS HAVE BEEN PLAYING FOR *YEARS.*

"REMEMBER BACK WHEN I HELPED YOU MAKE THOSE COSTUMES?

"AND YOU HAD ME BE AN ELF FOR YOUR BIG STORY FINALE?"

DOES THAT MEAN YOU DRESSED UP TOO?

YOU'VE WORKED SO HARD TO TELL THOSE STORIES.

"ARE YOU GOING TO LET SOME STUPID MONSTER TAKE THAT AWAY FROM YOU?"

"YOU GOT YOUR FRIEND BACK. MAKE SURE YOU GET YOUR NORMAL LIFE BACK TOO."

I MEAN, WILL'S NORMAL LIFE.

YEAH. THANKS, NANCY.

THE END

Dark Horse Direct Exclusive
Art by BELLA GRACE

Art by CLAIRE ROE

Art by EVAN CAGLE

Art by JONATHAN CASE

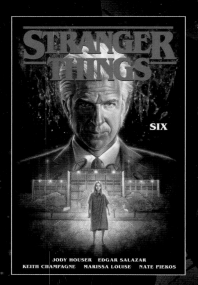